Lecture Notes in Mathematics

Edited by A. Dold and B. Eckmann

Subseries: Fondazione C.I.M.E., Firenze
Adviser: Roberto Conti

1225

Inverse Problems

Lectures given at the 1st 1986 Session of the
Centro Internazionale Matematico Estivo (C.I.M.E.)
held at Montecatini Terme, Italy, May 28 – June 5, 1986

Edited by G. Talenti

T0172222

Springer-Verlag
Berlin Heidelberg New York London Paris Tokyo

Editor

Giorgio Talenti
Istituto Matematico "U.DINI", Università di Firenze
Viale Morgagni 67/a, 50134 Firenze, Italy

Mathematics Subject Classification (1980): 31A25, 34A55, 35R25, 45B05

ISBN 3-540-17193-2 Springer-Verlag Berlin Heidelberg New York
ISBN 0-387-17193-2 Springer-Verlag New York Berlin Heidelberg

Printing and binding: Druckhaus Beltz, Hemsbach/Bergstr.
2146/3140-543210

INTRODUCTION

This volume collects the texts of the lectures and seminars delivered at the CIME Session on Inverse Problems (Montecatini Terme, Italy, May 28-June 5, 1986).

The Session was designed to present significant trends in the nowadays growing field of ill-posed and inverse problems, and consisted of the following courses, six lectures each: (i) Inverse eigenvalue problems for second-order and fourth-order Sturm-Liouville differential equations (V. Barcilon, University of Chicago); (ii) Fredholm integral equations of the first kind and regularization methods (M. Bertero, Università di Genova); (iii) Mathematical problems arising from tomography and related topics (A.F. Grunbaum, University of California, Berkeley); (iv) Numerical methods for large-scale ill-posed problems (F. Natterer, University of Munster). Furthermore, seminars were given on integral equations arising in optics (E.R. Pike, Royal Signals and Radar Establishment, Great Malvern, UK) and functional analysis arguments for classes of linear ill-posed problems (Ch. Groetsch, University of Cincinnati).

The editor is heartily grateful to all speakers for the excellent and stimulating atmosphere they were able to create in Montecatini, as well as for the accuracy of their notes, and thanks to CIME staff for its efficient job.

Firenze, July 1986

Giorgio Talenti

C.I.M.E. Session on "Inverse Problems"

List of Participants

V. ADOLFSSON, Matematiska Institutionen, CTH, 412 96 Goteborg, Sweden

G. ALESSANDRINI, Istituto Matematico Università, Viale Morgagni 67/A, 50134 Firenze

O. ARENA, Via dei Sette Santi 55, 50100 Firenze

V. BACCHELLI, Corso Sempione 48, 20154 Milano

V. BARCILON, The Univ. of Chicago, Department of the Geophysical Sciences, 5734 S. Ellis Avenue, Chicago, ILL. 60637, USA

E. BATTISTINI, Dipartimento di Matematica del Politecnico, Piazza L. da Vinci 32, 20133 Milano

E. BERETTA, Istituto di Analisi Globale e Applicazioni, Via di S. Marta 13/A, 50139 Firenze

M. BERTERO, Dipartimento di Matematica, Università, Via L.B. Alberti 4, 16132 Genova

P. BRIANZI, Dipartimento di Matematica, Università, Via L.B. Alberti 4, 16132 Genova

S. CAMPI, Istituto Matematico Università, Viale Morgagni 67/A, 50134 Firenze

M.R. CAPOBIANCO, Istituto per le Applicazioni della Matematica, Viale A. Gramsci 5, 80122 Napoli

F. DEL BUONO, Via Eden 15, 52010 Badia Prataglia (Arezzo)

C. DE MOL, Département de Mathématiques, Université Libre de Bruxelles, Campus Plaine CP 217, Bd du Triomphe, 1050 Bruxelles, Belgium

P. FIJALKOWSKI, Institute of Mathematics, Lodz University, ul. S. Banacha 22, 90-238 Lodz, Poland

C.W. GROETSCH, Department of Mathematical Sciences, University of Cincinnati, Old Chemistry Building, Cincinnati, Ohio 45221-0025, USA

A.F. GRUNBAUM, University of California at Berkeley, Department of Mathematics, Berkeley, CAL 94720, USA

G. INGLESE, Istituto di Analisi Globale e Applicazioni, Via di S. Marta 13/A, 50139 Firenze

H. LENNERSTAD, Askims Domarringsvag 3, 436 00 Askim, Sweden

M. LONGINETTI, Istituto di Analisi Globale e Applicazioni, Via di S. Marta 13/A, 50139 Firenze

C.M. MADERNA, Viale F.lli Cervi, Residenza Spiga, 20090 Segrate (Milano)

P. MANSELLI, Istituto Matematico Università, Viale Morgagni 67/A, 50134 Firenze

M. MILLER, 1915 1/2 Addison St. 203, Berkeley, CA 94704, USA

F. NATTERER, Westfalische Wilhelms-Universitat Munster, Inst. fur Numerische und Instrumentelle Mathematik, Einsteinstrasse 62, 4400 Munster, BRD

C. PAGANI, Via Clemente Sala 16, 20092 Cinisello Balsamo (Milano)

G. PAPI, Istituto Matematico Università, Viale Morgagni 67/A, 50134 Firenze

A. PASQUALI, Istituto Matematico Università, Viale Morgagni 67/A, 50134 Firenze

C. PUCCI, Istituto di Analisi Globale e Applicazioni, Via di S. Marta 13/A, 50139 Firenze

E.R. PYKE, Royal Signals and Radar Establishment, St. Andrews Road, Great Malvern, WR14 3PS, England

L. ROSSI COSTA, Dipartimento di Matematica del Politecnico, Piazza L. da Vinci 32, 20133 Milano

S. ROVIDA, Istituto di Analisi Numerica del CNR, Corso Carlo Alberto 5, 27100 Pavia

F. SACERDOTE, Dipartimento di Matematica, Università, Via Buonarroti 2, 56100 Pisa

G. SCHMIDT, Department of Mathematics and Statistics, McGill University, 805 Sherbrooke St. West, Montreal, Quebec H3A 2K6, Canada

K. SEIP, Department of Mathematics, The University of Trondheim, 7034 Trondheim, Norway

G. TALENTI, Istituto Matematico Università, Viale Morgagni 67/A, 50134 Firenze

A. VENTURI, Via Paisiello 15, 50100 Firenze

M. VERRI, Dipartimento di Matematica del Politecnico, Piazza L. da Vinci 32, 20133 Milano

S. VESSELLA, Istituto di Analisi Globale e Applicazioni, Via di S. Marta 13/A, 50139 Firenze

A. VOLCIC, Istituto di Matematica Applicata, Facoltà di Ingegneria, Università, 34100 Trieste

G. WEILL, Département de Mathématiques, Université de Tours, Faculté des Sciences, Parc de Grandmont, 37200 Tours, France

F. ZIRILLI, Via Grossi Gondi 43, 00162 Roma

TABLE OF CONTENTS

INVERSE EIGENVALUE PROBLEMS

Victor Barcilon

Department of the Geophysical Sciences

University of Chicago

Chicago, Illinois USA

I. INTRODUCTION AND PRELIMINARIES

1. Inverse problems in Geophysics

These lectures are a very biased and personal tour of Inverse Eigenvalue Problems. I intend to related what motivated my work in this subject, the techniques which I personally found useful, and my current understanding of the subject mátter. The lectures are certaintly not meant to be a comprehensive review of the field.

Inverse problems have always played a very important role in Geophysics. Relying on well-understood physical laws, geophysicts have traditionally looked upon the Earth as a black box which produces measurable outputs to various naturally applied inputs. Their task has been to infer the properties of the black box from measurements of these outputs. For instance, making use of the theory of thermal conduction as well as of measurements of heat fluxes at the surface of the Earth, Kelvin attempted to infer the thermal history of the Earth and in so doing its age (see Richter 1986 for a modern assesment of Kelvin's work). I mention this example for two reasons: first, because Kelvin got the wrong result. The reason being that neither convection nor radioactive heat sources were taken into account. This failure was due to the wrong Physics and is not a shortcoming of the inverse method which concerns itself solely with a mathematical problem. The second reason for mentioning this example is because not all inverse problems are associated with wave propagation.

Nevertheless, a great many inverse problems use propagating waves to probe regions otherwise inaccessible. Seismic waves, either naturally generated or brought about by a thumper, have been used extensily to learn about the structure of the Earth in the large and in the small. Inspite of routine use of these seismic waves, the geophysical world got very excited in 1959 when normal modes of vibrations of the Earth as a whole were recorded for the first time by Benioff et al. (1961) as well as Ness et al. (1961) and confirmed theoretically by Alterman et al.(1959). As with many discoveries, this observation was fortuitous. Indeed, previously it was thought doubtful that free seismic waves could travel around the Earth several times without attenuation and organize themselves into normal modes. This observation ushered in a new class of inverse problems in Geophysics, namely Inverse Eigenvalue Problems. Given a sufficient number of measured natural frequencies of vibration of the Earth, can we improve upon our knowledge of its interior?

Even with drastic oversimplifications, the mathematical statement of the above question is frightfully difficult. For instance, if we assume (i) that the Earth is a stationary sphere rather than a rotating ellipsoid, (ii) that its internal properties depend solely on the distance from the center, (iii) that gravitational effects are negligible and (iv) that there is no attenuation, we are forced to deal with the following two coupled inverse eigenvalue problems:

$$-S_{\ell n}^{2}\rho U_{\ell n} = r^{-2}\frac{d}{dr}\left[r^{2}(\lambda+2\mu)\frac{dU_{\ell n}}{dr}\right]$$
$$-r^{-2}\left[\ell(\ell+1)\mu + 2(\lambda+2\mu) - 2r\frac{d\lambda}{dr}\right]U_{\ell n}$$
$$-r^{-1}\ell(\ell+1)\left[\frac{d(\lambda\Psi_{\ell n})}{dr} + \mu\frac{d\Psi_{\ell n}}{dr} - r^{-1}(\lambda+3\mu)\Psi_{\ell n}\right],$$

$$-S_{\ell n}^{2}\rho\Psi_{\ell n} = r^{-2}\frac{d}{dr}\left[r^{2}\mu\frac{d\Psi_{\ell n}}{dr}\right] - r^{-2}\left[\ell(\ell+1)(\lambda+2\mu)\right.$$
$$\left. + r\frac{d\mu}{dr}\right]\Psi_{\ell n} + r^{-1}\frac{d(\mu U_{\ell n})}{dr} + r^{-1}\lambda\frac{dU_{\ell n}}{dr} + r^{-2}(\lambda+2\mu)U_{\ell n},$$

$$-T_{\ell n}^{2}\rho\Omega_{\ell n} = r^{-2}\frac{d}{dr}\left[r^{2}\mu\frac{d\Omega_{\ell n}}{dr}\right] - r^{-2}\left[\ell(\ell+1)\mu + r\frac{d\mu}{dr}\right]\Omega_{\ell n}.$$

I shall not bother to record the boundary conditions which express the fact that the Earth surface is stress free. Let me simply say that because of the assumed spherical symmetry, the normal modes fall into two classes: the spheroidal modes with frequencies $S_{\ell n}$ given by the first two equations and the toroidal modes with frequencies $T_{\ell n}$ given by the last equation. The eigenfunctions $U_{\ell n}$, $\Psi_{\ell n}$ and $\Omega_{\ell n}$ are related to the displacements in a spherical coordinate system. As might have been guessed, ℓ and n are angular and radial modal numbers. Also as is usually the case, the ℓ-th problem has a degeneracy of order $2\ell+1$. Finally, ρ, λ, μ are the density and Lame parameters: they are the unknown functions of r which we want to infer. Therefore, eventhough the spheroidal modes decouple from the toroidal ones in the direct version of the eigenvalue problem, in the inverse version these modes are coupled since ρ and μ appear in both.

In view of the intrinsic difficulties of the problem, we must lower our expectation and change slightly our goal. One approach, would be not to discard completely information obtained by other means, say travel times, and use solely the eigenvalues to refine this information. More precisely, this approach consists is postulating a-priori an Earth model about which to linearize the nonlinear inverse problem. This is essentially the approach taken by Backus & Gilbert (1967,1968). I say "essentially" because they went one important step further, namely they built into their analysis right from the start the fact that the data were finite in number and contaminated by errors. They were naturally led to consider the following technical question:

Given mismatches δS_p and δT_p between the measured spheroidal and toroidal frequencies and their counterparts as obtained from the Earth model, find corrections $\delta\rho$, $\delta\lambda$ and $\delta\mu$ such that:

$$\begin{bmatrix} \delta S_p \\ \\ \delta T_p \end{bmatrix} = \int \begin{bmatrix} G_{11}^{(p)}(r) & G_{12}^{(p)}(r) & G_{13}^{(p)}(r) \\ \\ G_{21}^{(p)}(r) & G_{22}^{(p)}(r) & G_{23}^{(p)}(r) \end{bmatrix} \begin{bmatrix} \delta\rho(r) \\ \delta\lambda(r) \\ \delta\mu(r) \end{bmatrix} dr$$

where p is an enumeration of the finite number of modes in the data set. Backus & Gilbert (1967) define and construct an inverse to the above integral equation. I shall not say more about their approach other than that it is reminiscent of the theory of generalized inverses of rectangular matrices. However, if the radial coordinate is discretized and the problem is truly reduced to the inversion of a rectangular matrix, their approach leads to a different inverse than the standard Moore-Penrose one.

Another approach consists in studying simpler eigenvalue problems in the hope of forging tools and developping insights about the more complicated problem of interest. This is the approach that I have followed and that I shall develop in these lectures. In particular, I shall devote all of my time to present results about the inverse problems for a vibrating string and for a vibrating beam.

2. The vibrating string

It has become traditional among workers on inverse eigenvalue problems to write the governing equation for a vibrating string thus:

$$-\omega_n^2 \rho u_n = \frac{d^2 u_n}{dx^2}, \qquad x \in (0,L)$$

In this equation, $u_n(x)$ is the shape of the n-th mode whose frequency is ω_n. Implicit in the above equation is the assumption that the tension necessary for vibrations to exist is constant. Eventhough $\rho(x)$ has units of an inverse square velocity, it is customary to refer to it as the density. I shall do so in the sequel. Finally, the string has length L.

The similarity between the equation for the vibrating string and the equations for the vibrating elastic sphere is obvious. There are differences too. Nevertheless, the problem of inferring the density of a vibrating string from spectral data is bound to have implications about the geophysical one.

Since the eigenvalues are all (or part) of the spectral data on hand, we should first ask what information is contained in these eigenvalues. A partial answer is obtained by considering their asympotic form. I shall not go through the detailed asymptotic analysis but simply remind you of some of the key steps. The method is that of WKBJ and consists in looking for a solution of the form

$$u_n = \cos(\omega_n \xi(x)) \, [\, g^{(0)}(x) + \omega_n^{-1} g^{(1)}(x) + \ldots \,]$$

$$+ \sin(\omega_n \xi(x)) \, [\, h^{(0)}(x) + \omega_n^{-1} h^{(1)}(x) + \ldots \,]$$

The function $\xi(x)$ entering in the phase is easily found, viz.

$$\xi(x) = \int_0^x \rho^{1/2}(t)\, dt,$$

whereas the amplitudes $g^{(0)}$ and $h^{(0)}$ turn out to be

$$g^{(0)}(x) = A^{(0)} \rho^{-1/4}(x),$$

$$h^{(0)}(x) = B^{(0)} \rho^{-1/4}(x).$$

The determination of the constants $A^{(0)}$ and $B^{(0)}$, as well as of ω_n, depends on the precise nature of the boundary conditions at the end points.

The most general boundary conditions are:

$$\cos\alpha \; u_n - L \sin\alpha \; du_n/dx = 0, \text{ at } x = 0,$$

and

$$\cos\beta \; u_n + L \sin\beta \; du_n/dx = 0, \text{ at } x = L.$$

where $\alpha, \beta \in [0,\pi)$.

We can think of a boundary condition as a point with coordinates (α,β) in the square $[0,\pi)\times[0,\pi)$. The edge $\alpha=0$ corresponds to the case in which the left end of the string is fixed. Similarly, the edge $\beta=0$ corresponds to a fixed right end. The corner $\alpha=\beta=0$ corresponds to a string with both end fixed. Finally, the point $\alpha=\beta=\pi/2$ which lies in the middle of the square corresponds to a string with both ends free (see Poeschel & Trubowitz 1984 for extensive geometric analysis of the role of boundary conditions).

We can now state the results of the asympotic analysis.

For $\alpha=\beta=0$, denoting the eigenvalues by λ_n, we have

$$\lambda_n^2 = n^2\pi^2/T^2 + O(1)$$

For either $\alpha=0$ or $\beta=0$, but not both, denoting the eigenvalues by μ_n, we have

$$\mu_n^2 = (n-1/2)^2\pi^2/\, T^2 + O(1).$$

Elsewhere in the square, we have

$$\omega_n^2 = (n-1)^2\pi^2/T^2 + O(1).$$

In the above formulas,

$$T = \int_0^L \rho^{1/2}(t)\,dt$$

which is the time for a disturbance to travel the length of the string.

The above results show that to leading order the eigenvalues contain <u>no</u> information on the density, other than the travel time T. Hence, they give an early warning of the difficulties associated with the recovery of ρ from a knowledge of the spectrum.

3. The Liouville transformation and the Sturm-Liouville problem

Let us push the asymptotic expansion to the next order of approximation. In order to compute $g^{(0)}$ and $h^{(0)}$ we must require that ρ be (i) twice differentiable and (ii) bounded away from zero. The actual algebra is simplified if we change variables thus:

$$\xi = \int_0^x \rho^{1/2}(t)\,dt,$$

$$U_n(\xi) = \rho^{1/4}(x)\,u_n(x).$$

This is the Liouville transformation. It changes the eigenvalue problem for the vibrating string to the classical Sturm-Liouville problem, viz.

$$d^2U_n/d\xi^2 + (\omega_n^2 - q(\xi))\,U_n = 0, \qquad \xi \in (0,T),$$

$$\cos a\, U_n - T \sin a\, dU_n/d\xi = 0 \text{ at } \xi = 0,$$

$$\cos b\, U_n + T \sin b\, dU_n/d\xi = 0 \text{ at } \xi = T.$$

The correspondence between the various quantities is:

$$q(\xi) = \frac{\rho_{xx}}{4\rho} - \frac{5\rho_x^2}{16\rho^3},$$

$$\frac{\cot a}{T} = \frac{\cot\alpha}{L\rho^{1/2}(0)} + \frac{\rho_x(0)}{4\rho^{3/2}(0)},$$

$$\frac{\cot b}{T} = \frac{\cot\beta}{L\rho^{1/2}(L)} - \frac{\rho_x(L)}{4\rho^{3/2}(L)}.$$

The derivation of the asymptotic form of ω_n as a functional of $q(\xi)$ though still difficult is nevertheless much easier than as a functional of $\rho(x)$. Postponing for the time being a discussion of the inverse Liouville transformation, we state the asympotic results thus (see e.g. Borg 1946, Hochstad 1961).

For $q \in L(0,T)$ and for $\alpha = \beta = 0$,

$$\lambda_n^2 = n^2\pi^2T^{-2} + \bar{q} - 1/2\, q_{2n+2} + o(1/n).$$

For either $\alpha = 0$ or $\beta = 0$ but not both,

$$\mu_n^2 = (n-1/2)^2\pi^2T^{-2} + \bar{q} + 2\,T^{-2}\cot c - 1/2\, q_{2n+1} + o(1/n),$$

where c stands for either a or b, whichever is nonzero;

For both a and b different from zero

$$\omega_n^2 = (n-1)^2\pi^2T^{-2} + \bar{q} + 2\,T^{-2}(\cot a + \cot b) + 1/2\, q_{2n} + o(1/n).$$

In the above formulas

$$\bar{q} = T^{-1} \int_0^T q(\xi)\, d\xi ,$$

and

$$q_n = 2T^{-1} \int_0^T q(\xi) \cos n\pi\xi/T \, d\xi .$$

The above asymptotic formulas suggest that a spectrum contains roughly speaking at least half of the Fourier coefficients of q. Could we extract more information by going to the next order of approximation? We shall shortly see that the answer is no. We shall also make rigorous this suggestion that two spectra are required to infer q.

Several questions peripherally related to our main topic have been left dangling. In particular, what does the determination of $q(\xi)$ have to do with that of $\rho(x)$? Does the Liouville transformation have a unique inverse? Also, what can we do, if anything, if $\rho(x)$ is not smooth and is in fact discontinuous, as we suspect is the case for the Earth?

The inverse Liouville transformation does not uniquely specify ρ. Two extra pieces of data are required for that purpose.

Discontinuities in ρ manifest themselves in the eigenvalues. This question has been discussed by Anderssen (1977), Hald (1984) and by Willis (1985).

4. Gelfand-Levitan formulas

I close this lecture by stating without proof three intriguing formulas first derived by Gelfand & Levitan (1953):

$$\sum_1^{\infty} (\bar{\lambda}_n^2 - \lambda_n^2) = \frac{1}{4} \{ q(0) + q(T) - 2\bar{q} \},$$

$$\sum_1^{\infty} (\bar{\mu}_n^2 - \mu_n^2) = \frac{1}{4} \{ - q(0) + q(T) \},$$

$$\sum_1^{\infty} (\bar{\omega}_n^2 - \omega_n^2) = \frac{1}{4} \{ - q(0) - q(T) + 2\bar{q} \}.$$

In these formulas, λ_n, μ_n and ω_n are eigenvalues corresponding to various boundary conditions. The overbar indicates that the eigenvalue is associated with a constant potential equal to the average value of q, namely $\bar{q}$. In some sense, these formulas are the inverse of the asymptotic ones. Indeed, while the asymptotic formulas provide us with a peek at the functional dependence of the eigenvalues on q(x), these formulas offer a glimpse of the dependence of q on the eigenvalues. This observation is the starting point of a construction procedure which I shall sketch in the sequel.

II. INVERSE STURM-LIOUVILLE PROBLEM

1. Uniqueness theorem: historical overview

Recall the basic problem:

$$U_{\xi\xi} + (\omega^2 - q)U = 0, \qquad \xi \in (0,T),$$

with

$$\cos a \; U - T \sin a \; U_\xi = 0 \quad \text{at } \xi = 0,$$

$$\cos b \; U + T \sin b \; U_\xi = 0 \quad \text{at } \xi = T.$$

The first uniqueness result is due to Borg(1946). He proved that given two spectra associated with particular choices of the boundary conditions, if q exists it is unique as a function in $L_2(0,T)$. If $\Omega(a,b)$ denotes the spectrum associated with the boundary conditions specified by a and b, then Borg's result is as follows:

Given a_1, a_2 and b such that

$$|\sin a_1| + |\sin a_2| \neq 0,$$

$$|\sin a_1| \; |\sin a_2| = 0,$$

and $\Omega(a_1,b)$, $\Omega(a_2,b)$, then if $q(\xi)$ exists it is unique in $L_2(0,T)$.

Thus, in the square $[0,\pi)\times[0,\pi)$ the points (a_1,b) and (a_2,b) must lie on a horizontal line, i.e. the right boundary condition is the same for both problems. More importantly, the two points must be in either one of two configurations:

(i) $(0,0)$ and $(a_2,0)$

or

(ii) $(0,b)$ and (a_2,b).

These choices are easily understood in the light of the asymptotic formulas given earlier.

In order to understand the method used by Borg, suppose that q is changed by δq. Then the eigenvalues are changed by a small amount $\delta\omega_n$ such that

$$\delta\omega_n^2 = \int_0^T \delta q \; U_n^2 \, d\xi$$

If the spectra $\Omega(a_1,b)$ and $\Omega(a_2,b)$ define q uniquely, then it must follow that the <u>square</u> of the eigenfunctions associated with two spectra taken together form a complete set. Borg established the uniqueness theorem by proving the completeness of squares of eigenfunction sets.

Levinson's (1949) contribution is twofold: he greatly géneralized the admissible pairs of boundary conditions. In particular, any two spectra $\Omega(a_1,b)$ and $\Omega(a_2,b)$ such that $a_1 \neq a_2$ are acceptable. Furthermore, the proof relies on the fact that solutions of the differential equation are entire functions of ω^2. This observation holds true even for higher order problems and hence this method can easily be generalized.

Finally, Marchenko (1952) has removed the parameters a_1, a_2, b and T from the data set. They can also be inferred from the eigenvalues.

2. Levinson's proof

__Theorem:__ Given two spectra $\Omega(a_1,b)$, $\Omega(a_2,b)$ such that $a_1 \neq a_2$, then q is uniquely determined.

Proof: We shall assume that two potentials q and $\tilde{q}$ give rise to the same pair of spectra.
Let $Y(\xi, \omega^2)$ be the solution of the initial value problem:

$$Y_{\xi\xi} + (\omega^2 - q)Y = 0,$$

$$Y(T,\omega^2) = \sin b,$$

$$Y_\xi(T,\omega^2) = - T^{-1}\cos b.$$

Similarly, let $U(\xi,\omega^2)$ be the solution of the same equation which satisfies the initial conditions

$$U(0,\omega^2) = \sin a_1$$

$$U_\xi(0,\omega^2) = T^{-1}\cos a_1.$$

We know that Y and U are independent solutions provided that the wronskian

$$W(\omega^2)= \begin{vmatrix} U & Y \\ U_\xi & Y_\xi \end{vmatrix}$$

is different from zero. In the above expression for W we have tacitly indicated that the wronskian is solely a function of ω^2. As a result, we can evaluate the entries in the determinant at $\xi=0$ and write

$$W(\omega^2) = \sin a_1 \, Y_\xi(0,\omega^2) - T^{-1}\cos a_1 \, Y(0,\omega^2).$$

This shows that $W(\omega^2)$ vanishes for $\omega^2 \in \Omega(a_1,b) = \{\omega_n^{(1)2}\}_1^\infty$. As a result, for $\omega = \omega_n^{(1)}$ the solutions Y and U are linearly related:

$$U(\xi,\omega_n^{(1)2}) = C_n\, Y(\xi,\omega_n^{(1)2}),$$

$$U_\xi(\xi,\omega_n^{(1)2}) = C_n\, Y_\xi(\xi,\omega_n^{(1)2}).$$

The constants of proportionality which will play an important role in the sequel, can be evaluated explicitly. This is accomplished by first recalling that both U and Y, and hence W, are entire functions of ω^2 (see Hille 1968). Furthermore, these entire functions are of order 1/2 (see Boas 1954 for a definition of order as well as for an excellent introduction to the theory of entire functions). As a result they admit very simple infinite product representations. In particular,

$$\sin a_2\, Y_\xi(0,\omega^2) - T^{-1}\cos a_2\, Y(0,\omega^2) = K\prod_n^\infty(1 - \omega^2/\omega_n^{(2)2})$$

The constant K is also known explicitly. Its value is obtained by considering the asymptotic form of the above equation. Indeed, for large ω^2, the leading terms of Y and Y_ξ, being independent of q, are completely known. Setting $\omega^2 = \omega_m^{(1)2}$, we deduce that

$$C_m^{-1}\sin(a_1 - a_2) = K\prod_n^\infty(1 - \omega_m^{(1)2}/\omega_n^{(2)2}).$$

As a result, if we also define corresponding solutions $\tilde{U}$ and $\tilde{Y}$ associated with $\tilde{q}$, then

$$\tilde{U}(\xi,\omega_n^{(1)2}) = C_n\, \tilde{Y}(\xi,\omega_n^{(1)2}),$$

$$\tilde{U}_\xi(\xi,\omega_n^{(1)2}) = C_n\, \tilde{Y}_\xi(\xi,\omega_n^{(1)2}),$$

where the constants of proportionality are the same as before.

The next step consists in defining a Cauchy integral analogous to that which arises in the classical proof of the completeness of the eigenfunctions of the Sturm-Liouville problem (Titchmarch 1948),viz.

$$\tilde{I}(\xi) = -\frac{1}{2\pi i}\int_0^\xi h(\zeta)\,d\zeta\int_{-i\infty}^{i\infty} ds\,\frac{\tilde{U}(\zeta,-s)\,Y(\xi,-s)}{W(-s)}$$

$$-\frac{1}{2\pi i}\int_\xi^T h(\zeta)\,d\zeta\int_{-i\infty}^{i\infty} ds\,\frac{U(\xi,-s)\,\tilde{Y}(\zeta,-s)}{W(-s)},$$

where $h(\xi)$ is an arbitrary function vanishing at ξ=0,T. The function $h(\xi)$ can be thought of as the arbitrary shape of a string released from rest. Then $\tilde{I}(\xi)$, or rather $I(\xi)$ which is similarly defined except for the fact that tildes are deleted, is the inverse Laplace transform of the corresponding initial value problem evaluated at time t=0.

We evaluate $\tilde{I}(\xi)$ in two different ways. First, by the calculus of residues. After closing the contour integrals in the left hand plane, the enclosed poles are the zeros of the wronskian, namely

$$s = -\omega_n^{(1)2}, \qquad n = 1,2,...$$

As a result

$$I(\xi) = -\sum_1^{\infty} \Big[\frac{C_n}{\dot{W}_n} \int_0^T h(\zeta)\, \tilde{U}_n^{(1)}(\zeta)\, d\zeta \Big] U_n^{(1)}(\xi),$$

where

$$U_n^{(1)}(\xi) = U(\xi,\omega_n^{(1)2}),$$

$$\tilde{U}_n^{(1)}(\xi) = \tilde{U}(\xi,\omega_n^{(1)2}),$$

and

$$\dot{W}_n = W(\omega_n^{(1)2}).$$

The second way of evaluating $\tilde{I}$ is by replacing the integrand by the leading term in its asymptotic expansion in s. Straightforward calculations show that

$$\tilde{I}(\xi) = h(\xi).$$

Thus, we obtain the hybrid eigenfunction expansion:

$$h(\xi) = \sum_1^{\infty} \Big[-\frac{C_n}{\dot{W}_n} \int_0^T h(\zeta)\, \tilde{U}_n^{(1)}(\zeta)\, d\zeta \Big] U_n^{(1)}(\xi).$$

Of course, the classical eigenfunction expansion, namely

$$h(\xi) = \sum \Big[-\frac{C_n}{\dot{W}_n} \int_0^T h(\zeta)\, U_n^{(1)}(\zeta)\, d\zeta \Big] U_n^{(1)}(\xi),$$

can be obtained by repeating the same procedure on the integral $I(\xi)$ alluded to earlier. As a result

$$\int_0^T h(\zeta)\, [\tilde{U}_n^{(1)}(\zeta) - U_n^{(1)}(\zeta)]\, d\zeta = 0, \text{ for } n=1,2,...$$

Consequently,

$$\tilde{U}_n^{(1)} = U_n^{(1)},$$

and

$$\tilde{q}(\xi) = q(\xi) \qquad \text{a.e.}$$

3. Construction of solution

Several construction methods have appeared in the literature. A critique of their merits with regards to computational ease or generalization is still largely missing. The method which I shall sketch is selected purely on the basis of its novelty. It combines the Gelfand-Levitan formulas of the previous lecture with the function theoretic arguments of Levinson's proof.

For the sake of simplicity, let me restrict my attention to the case in which the given eigenvalues are $\{\lambda_n\}$ and $\{\mu_n\}$ corresponding to the boundary bonditions $U(0)=U(T)=0$ and $U'(0)=U(T)=0$ respectively. Then

$$q(0) = 2\sum_1^{\infty} \{ \mu_n^2 - \frac{(n-1/2)^2\pi^2}{T^2} - \lambda_n^2 - \frac{n^2\pi^2}{T^2} \}$$

If the interval is shrunk to (t,T), then the above formula would read

$$q(t) = 2\sum_1^{\infty} \{ \mu^2(t) - \frac{(n-1/2)^2\pi^2}{(T-t)^2} - \lambda^2(t) - \frac{n^2\pi^2}{(T-t)^2} \},$$

where $\{\lambda_n(t)\}$ and $\{\mu_n(t)\}$ are the eigenfunctions corresponding to the boundary conditions $U(t)=U(T)=0$ and $U'(t)=U(T)=0$ respectively. Can we infer these eigenvalues from the given ones ? The answer is yes, and this is done by exploiting the analytic dependence of $Y(\xi,\omega^2)$ on ω. I shall omit the intermediary steps and simply write the differential equations which must be solved:

$$\frac{d\lambda_n^2}{dt} = G(t)\ \lambda_n^2\ \frac{\prod_k (1-\lambda_n^2/\mu_k^2)}{\prod_{k\neq n} (1-\lambda_n^2/\lambda_k^2)},$$

$$\frac{d\mu_n^2}{dt} = \{\frac{\mu_n^4}{G(t)} - q(t)\ G(t)\ \mu_n^2\}\ \frac{\prod_k (1-\lambda_n^2/\mu_k^2)}{\prod_{k\neq n} (1-\mu_n^2/\mu_k^2)},$$

$$\frac{dG}{dt} = q(t) - G^2,$$

with initial conditions

$$\lambda_n(0) = \lambda_n,$$

$$\mu_n(0) = \mu_n,$$

and

$$G(0) = \lim_{\omega\to\infty} \frac{\tan \omega T}{\omega} \prod_{n=1}^{\infty} \frac{1 - \omega^2/\mu_n^2}{1 - \omega^2/\lambda_n^2} .$$

4. Concluding remarks

In order to motivate the topic of the next lecture, I would like to remind you that the transformation of the vibrating string into a Sturm-Liouville problem places severe restrictions on the class of allowable densities. Also, the need for two spectra is odd physically. Other equivalent but more readily available data should be considered. For these reasons we shall reconsider the problem for the string ab initio.

III. INVERSE PROBLEM FOR THE VIBRATING STRING

1. Impulse response

Following Krein (1952), let us consider the following initial value problem for the vibrating string:

$$\rho(x)\, y_{tt} = y_{xx} + \delta(t)\delta(x),$$

with initial conditions

$$y(x,0) = y_t(x,0) = 0,$$

and boundary conditions

$$y_x(0,t) = y(L,t) = 0,$$

in other words, let us consider a string with a free left end and a fixed right one, set in motion from rest by a point force applied at the left end at the origin of time. We should note at this stage the analogy between this model problem and the typical seismological problem consisting of calculating the motion of the Earth resulting from the occurence of an earthquake. In such a calculation, the earthquake is often represented by an instantaneous point disturbance; furthermore, because of the shallowness of the epicenters, the point disturbance is often assumed to be on the surface.

Let us consider that $\rho(x)$ is unknown and that the data consist of the time record of the displacement of the left end of the string. In the seismological case, this record is provided by a seismograph. We shall refer to the data $y(0,t)$ as the impulse response.

Given $y(0,t)$ can we infer $\rho(x)$?

Let us first convince ourselves that the knowledge of the impulse response is equivalent to the knowledge of two spectra with different left end boundary conditions but identical right ones. To that effect, we Laplace transform the problem. With

$$\hat{y}(x,s) = \int_0^\infty y(x,t) \exp(-st)\, dt,$$

the problem becomes

$$s^2\rho(x)\, \hat{y} = \hat{y}_{xx}$$

with

$$\hat{y}_x(0+,s) + 1 = \hat{y}(L,s) = 0.$$

If $u(x,\omega^2)$ is that fundamental solution of

$$u_{xx} + \omega^2 \rho(x) u = 0,$$

such that

$$u(L,\omega^2) = u_x(L,\omega^2) - 1 = 0,$$

then clearly

$$\hat{y}(x,s) = - \frac{u(x,-s^2)}{u_x(0,-s^2)} .$$

Since both u and u_x are entire functions of ω^2 of order 1/2, they admit the following infinite product representations:

$$u(x,\omega^2) = (x - L) \prod_n^{\infty} (1 - \omega^2 / \lambda_n^2(x)),$$

$$u_x(x,\omega^2) = \prod_n^{\infty} (1 - \omega^2 / \mu_n^2(x)).$$

where $\lambda_n(x)$ is the nth eigenvalue of that portion of the string from x to L fixed at both ends. Similarly, $\mu_n(x)$ is the nth eigenvalue for the same truncated string but with a free left end and a fixed right one. Thus, in frequency space, the impulse response is

$$\hat{y}(0,s) = L \frac{\prod_n^{\infty} (1 + s^2/\lambda_n^2)}{\prod_n^{\infty} (1 + s^2/\mu_n^2)} .$$

The Fourier transform of the impulse response is a meromorphic function whose zeros and poles correspond to two spectra associated with idendical right end boundary conditions but distinct left end ones.

Thus, the information content of the impulse response is equivalent to a two-spectra data set, by which we mean the knowledge of the length of the string and of two spectra associated with distinct left-end boundary conditions and identical right-end ones.

Another set of equivalent data is obtained by writing $\hat{y}(0,s)$ as a partial fraction. This alternative representation is obtained by expressing $\hat{y}(x,s)$ in terms of the eigenfunctions

$$u_n(x) = u(x,\mu_n^2),$$

of the string in the free-fixed vibrating configuration. Omitting the some simple calculations, we deduce that

$$\hat{y}(0,s) = \sum_n^{\infty} \frac{u_n^2(0)}{(s^2 + \mu_n^2) \int_0^L \rho u_n^2(x)\, dx} .$$

Thus the impulse response, and hence the two-spectra data set, is equivalent to a spectrum-end values data set. By spectrum-end values data set, we mean the knowledge of one spectrum and a combination of amplitude and slope of the associated normalized eigenfunctions at the left end point.

Incidentally, by confronting the two different expressions for the impulse response for s=0, we obtain a formula for evaluating the length of the string:

$$L = \sum_n^{\infty} \frac{u_n(0)}{\mu_n^2} ,$$

where the eigenfunctions are assumed to be normalized.

2. Krein discretization

Rather than discretizing the density, Krein discretizes the mass

$$M(x) = \int_0^x \rho(t)\, dt.$$

There is one great advantage to this approach: it leads to a system with a finite number of degrees of freedom and hence to a finite number of eigenvalues. Indeed, suppose that we are given two sequences $\{\lambda_n\}_1^{\infty}$ and $\{\mu_n\}_1^{\infty}$ and that we set ourselves the task of constructing that string whose first N pairs of eigenfrequencies in the fixed-fixed and free-fixed configurations coincide with the first N pairs of terms in the sequences. We can clearly look for such a string among those made up of N unknown point masses separated by massless segments of length. Krein gave an elegant construction of such a string.

We next outline Krein's construction (1952). Consider a discrete string made up of N point masses located at points $x_1, \ldots , x_N$ such that

$$l_i = x_{i+1} - x_i .$$

As in the continuum case considered earlier, the right end at $x_{N+1} = L$ is assumed to be fixed. If u_i represents the displacement of the ith point mass of the string as it undergoes periodic oscillations with frequency ω, then Newton's law implies that

$$- \omega^2 m_i u_i = \theta_i - \theta_{i-1} ,$$

where, on account of the smallness of the oscillation, the angle θ_i is given by

$$\theta_i = (u_{i+1} - u_i)\, l_i^{-1}.$$

The displacement u_0 at the left end can therefore be expressed as

$$-u_0/\theta_0 = l_1 + 1/(-\theta_0/u_1),$$

or since

$$-\theta_0/u_1 = -m_1\omega^2 + 1/(-u_1/\theta_1),$$

we can repeat this process and write

$$-u_0/\theta_0 = l_0 + \frac{1|}{|-m_1\omega^2} + \frac{1|}{|l_1} + \frac{1|}{|-m_2\omega^2} + \dots + \frac{1|}{|l_N}.$$

Thus, by writing $-u_0/\theta_0$, which is the discrete analog of the impulse response, as a continuous fraction we can read of the solution to the inverse problem. Krein has shown that for any two positive, finite sequences of numbers $\{\lambda_n\}_1^N$, $\{\mu_n\}_1^N$ and a positive constant L, there exists an N-string with positive point masses $\{m_i\}_1^N$ separated by positive intervals $\{l_i\}_0^N$ adding up to L provided that the two sequences interlace, i.e. provided that

$$\mu_1 < \lambda_1 < \mu_2 < \dots < \lambda_N$$

and that the total mass is finite. This last condition can be dispensed with. We shall not pursue this approach. Instead, we shall start from the one-spectrum/end-point data set and follow the treatment of Gladwell & Gbageyan (1985).

3. The Gladwell-Gbadeyan construction

As is appropriate for the spectrum/end-value data set, we shall now consider that the eigenvectors

$$\mathbf{u}_n^T = [u_{n,0} \quad u_{n,1} \quad \dots \quad u_{n,N-1}], \qquad n=1,2, \dots, N$$

have been normalized, namely

$$\mathbf{U}^T \mathbf{M} \mathbf{U} = \mathbf{I},$$

where as usual a superscript T denotes the transpose, $\mathbf{I}$ is the NxN unit matrix, $\mathbf{M}$ is a diagonal matrix with entries $(m_0, \dots, m_{N-1})$ and

$$\mathbf{U} = [\mathbf{u}_1 \quad \mathbf{u}_2 \qquad \mathbf{u}_N].$$

It should be noted that in order to stress the analogy with the continuum, we have switched the traditional order of the subscripts: the first index is a nodal number whereas the second denotes a spatial point. As a result

$$\mathbf{U} = \begin{vmatrix} u_{1,0} & \cdots & u_{N,0} \\ \cdot & \cdot\ \cdot\ \cdot\ \cdot\ \cdot\ \cdot & \cdot \\ u_{1,N-1} & \cdots & u_{N,N-1} \end{vmatrix}$$

This will not cause any problems.

Pursuing our restatement of the problem in matrix notations, we write Newton's law thus:

$$\mathbf{M}\,\mathbf{U}\,\boldsymbol{\Omega} = -\,\mathbf{E}^{T}\,\boldsymbol{\Theta},$$

whereas the geometry of the deformed string implies that

$$\boldsymbol{\Theta} = -\,\mathbf{L}^{-1}\,\mathbf{E}\,\mathbf{U}.$$

In the above equations, $\mathbf{L}$ stands for the diagonal matrix with entries $(l_0, l_1, \ldots, l_{N-1})$, $\boldsymbol{\Omega}$ stands for the diagonal matrix with entries $(\mu_1^2, \mu_2^2, \ldots, \mu_N^2)$,

$$\boldsymbol{\Theta} = [\,\boldsymbol{\Theta}_1, \boldsymbol{\Theta}_2, \ldots, \boldsymbol{\Theta}_N\,],$$

and finally

$$\mathbf{E} = \begin{vmatrix} 1 & -1 & 0 & \ldots & 0 \\ 0 & 1 & -1 & \ldots & 0 \\ \cdot & \cdot & \cdot & \cdots & \cdot \\ 0 & 0 & 0 & \ldots & -1 \\ 0 & 0 & 0 & \ldots & 1 \end{vmatrix}.$$

This discretization, though analogous in spirit to that used by Krein in his discussion of the two-spectra version of the problem, differs nevertheless from the latter. The differences can be traced to the fact that for the spectrum/end-value problem, the truncated data set consits of $\{u_{n,0}\}_1^N$ and $\{\omega_n\}_1^N$, i.e of 2N pieces of data instead of 2N+1 as in the two-spectra version. To accomodate this loss of information, the last link consisting of m_N and l_{N+1} is deleted and, instead, a mass m_0 is introduced at x_0.

The solution is obtained as follows. First, from the completeness of the eigenvectors, viz.

$$\mathbf{U}\,\mathbf{U}^{T} = \mathbf{M}^{-1},$$

we can infer the mass m_i from a knowledge of $\{u_{n,i}\}_{n=1}^{N}$:

$$m_i^{-1} = \sum_{n=1}^{N} u_{n,i}^2, \qquad i=0,1, \ldots ,N-1.$$

In particular, m_0 is easily deduced from the data. Next, $\{\Theta_{n,i}\}_{n=1}^{N}$ are computed:

$$\Theta_{n,i} = \Theta_{n,i-1} - m_i \mu_n^2 u_{n,i}.$$

In particular,

$$\Theta_{n,0} = - m_0 \mu_n^2 u_{n,0}.$$

Thirdly, the length l_i of the interval is found by exploiting once again the completeness relation:

$$l_i = - \sum_{n=1}^{N} u_{n,i}^2 \Big/ \sum_{n=1}^{N} u_{n,i}\Theta_{n,i}.$$

In particular, l_0 can be found on the first go around. Finally, $\{u_{n,i+1}\}_{n=1}^{N}$ are obtained from the geometry of the string:

$$u_{n,i+1} = u_{n,i} + l_i \Theta_{n,i}.$$

The cycle is complete. Two questions remain: are the masses and intervals thus obtained positive and (ii) does this procedure converge as $N \to \infty$. We discuss these questions next.

4. Existence of a solution

The positivity of the masses is guaranteed by the formula used for their evaluation. Therefore, we focus our attention on the positivity of the intervals.

If we eliminate $\mathbf{U}$ from the problem, then $\boldsymbol{\Theta}$ satisfies

$$\mathbf{B}\boldsymbol{\Theta} = \mathbf{L}\boldsymbol{\Theta}\boldsymbol{\Omega}$$

where

$$\mathbf{B} = \mathbf{E}\,\mathbf{M}^{-1}\,\mathbf{E}^T$$

is a real symmetric matrix. It is easy to show that the orthogonality condition reads

$$\boldsymbol{\Theta}^T \mathbf{L}\boldsymbol{\Theta} = \mathbf{I},$$

and that the completeness relation is

$$\boldsymbol{\Theta}\,\boldsymbol{\Theta}^T = \mathbf{L}^{-1}.$$

From the above relation we deduce an alternative formula for l_i, namely

$$l_i^{-1} = \sum_{n=1}^{N} \theta_{n,i}^2,$$

which guarantees the positivity of l_i. In the next lecture, I shall return to this question and give an alternative way of establishing the existence of a solution.

5. The continuum limit

The eigenvalue problem for $\mathbf{U}$ is:

$$\mathbf{A}\,\mathbf{U} = \mathbf{M}\,\mathbf{U}\,\boldsymbol{\Omega},$$

where

$$\mathbf{A} = \mathbf{E}^T\,\mathbf{L}^{-1}\,\mathbf{E}.$$

This form of the eigenvalue problem is analogous to the differential equation form of the continuum problem. We want to transform it to the form which is analogous to the integral equation version, namely

$$\mathbf{U}\,\boldsymbol{\Omega}^{-1} = \mathbf{A}^{-1}\,\mathbf{M}\,\mathbf{U}.$$

Post-multiplying by $\mathbf{U}^T$ and using the completeness relation we deduce that

U

$$\begin{aligned}\mathbf{U}\,\boldsymbol{\Omega}^{-1}\mathbf{U}^T &= \mathbf{A}^{-1},\\ &= (\mathbf{E}^T\,\mathbf{L}^{-1}\,\mathbf{E})^{-1},\\ &= \mathbf{E}^{-1}\,\mathbf{L}\,\mathbf{E}^{-T},\end{aligned}$$

and by focusing on the 1-1 entries

$$\sum_{n=1}^{N} \mu_n^{-2}\, u_{n,0}^2 = \sum_{k=0}^{N-1} l_k^{(N)}.$$

This is the first gross relation between the data and the solution. Provided that the data satisfy the asymptotic trends, it implies that the total lengths of the discrete strings tend to L.
A second gross relation is obtained by equating the trace of $\boldsymbol{\Omega}^{-1}$ with that of $\mathbf{A}^{-1}\,\mathbf{M}$, namely

$$\sum_{n=1}^{N} \mu_n^{-2} = \sum_{i=0}^{N-1} m_i \sum_{k=i}^{N-1} l_k.$$

We make use of this relation as follows. First, we define the functions

$$M^{(N)}(x) = \sum_{k=1}^{i} m_k^{(N)} \qquad \text{for } x \in [x_i^{(N)}, x_{i+1}^{(N)}).$$

Clearly, $\{M^{(N)}(x)\}$ is a sequence of positive, nondecresing functions defined on $(0,x_N^{(N)}) \in [0,L]$. Next, we define their integrals

$$P^{(N)}(x) = \int_0^x M^{(N)}(t)\, dt.$$

It is equally clear that $\{P^{(N)}(x)\}$ is a sequence of positive, incresing functions. Furthermore, these functions are bounded above. Indeed, after some simple rearrangements, we can see that

$$P^{(N)}(x_N^{(N)}) = \sum_{k=0}^{N-1} m_k^{(N)} \sum_{j=k}^{N-1}{}' \; l_j^{(N)} ,$$

and as a result of the trace relation

$$P^{(N)}(x) \leqq \sum_{n=1}^{\infty} \mu_n^{-2} .$$

This in turn implies that $M^{(N)}(x)$ is bounded. In order to see this we write

$$\int_x^{x_N^{(N)}} M^{(N)}(t)\, dt = P^{(N)}(x_N^{(N)}) - P^{(N)}(x),$$

which, because of the monotonicity of $M^{(N)}(x)$ and $P^{(N)}(x)$, can be written as

$$M^{(N)}(x) < \frac{2P^{(N)}(x_N^{(N)})}{x_N^{(N)} - x} .$$

Therefore, provided we stay away from the right end x=L of the string, the sequence $\{M^{(N)}(x)\}$ is bounded above. Consequently, by Helly's selection principle (see e.g. Natanson 1955), there exists a subsequence which converges to a positive function M(x), which is the integrated mass.

IV. OSCILLATING AND TOTALLY POSITIVE OPERATORS

1. Totally positive matrices

The matrix **A** which we have encountered in our discussion of the discrete version of the vibrating string possesses some remarkable properties. These properties are shared by other matrices, and in fact by operators, arising among other places in the study of vibration problems. I am referring to the property of "total positivity" and to the closely related one of "oscillatoriness". A good introduction to the theory of totally positive matrices and operators can be found in Gantmakher & Krein (1937, 1950) and Karlin (1968). Here, we shall briefly touch upon this theory and list those results which will be necessary for a treatment of the vibrating beam.

Consider a generic NxN matrix **A**, where as usual

$$\mathbf{A} = \begin{vmatrix} a_{11} & \cdots & a_{1N} \\ \cdot & & \cdot \\ \cdot & & \cdot \\ \cdot & & \cdot \\ a_{N1} & \cdots & a_{NN} \end{vmatrix}$$

We shall denote by $A\begin{pmatrix} i_1 & i_2 & \cdots & i_p \\ j_1 & j_2 & \cdots & j_p \end{pmatrix}$ the determinant of that matrix formed by retaining the rows $i_1, i_2, \ldots, i_p$ and the columns $j_1, j_2, \ldots, j_p$. Such a determinant is called a minor. The matrix **A** is said to be <u>totally positive</u> (or completely positive) if all the minors of **A** are positive.

Clearly, all the entries of a totally positive matrix are positive. Hence totally positive matrices are necessarily <u>positive matrices</u>. Furthermore, totally positive matrices are also <u>positive definite matrices</u>, since

$$A\begin{pmatrix} 1 \\ 1 \end{pmatrix} > 0, \quad A\begin{pmatrix} 1 & 2 \\ 1 & 2 \end{pmatrix} > 0, \text{ etc}, \ldots$$

Similarly, we can define totally negative, totally non-positive and totally non-negative matrices.

As Gantmakher & Krein remark, the set of totally non-negative matrices is too large to be useful while the set of totally positive matrices is too small. Oscillatory matrices fit somewhere in between these two sets.

By definition, a totally non-negative matrix **A** is said to be <u>oscillatory</u> (or oscillating) if there exists an integer K such that

$\mathbf{A}^K$ is totally positive. Gantmakher & Krein (1950,p. 139) prove that a totally non-negative matrix **A** is oscillatory if it is (i) non-singular and such that (ii) the quasi-principal elements $a_{i,i+1}$ and $a_{i+1,i}$ are positive.

What are the properties of the inverse of a totally positive matrix **A**? Clearly $\mathbf{A}^{-1}$ cannot be totally positive. Indeed, if $\mathbf{B} = \mathbf{A}^{-1}$, then

$$b_{ij} = (-1)^{i+j} \frac{A\begin{pmatrix} 1 & \dots & i-1 & i+1 & \dots & N \\ 1 & \dots & j-1 & j+1 & \dots & N \end{pmatrix}}{A\begin{pmatrix} 1 & \dots & N \\ 1 & \dots & N \end{pmatrix}}$$

is negative for odd values of i+j. However, the matrix $\mathbf{B}^*$ defined thus

$$b^*_{ij} = (-1)^{i+j}\, b_{ij}$$

is totally positive.

The star operation associates to every totally non-negative (positive) matrix another matrix which is called sign-regular (strictly sign-regular). The inverse of a totally positive matrix is a strictly sign-regular matrix. The matrices encountered in the previous lecture provide us with examples of most of the matrices we have just defined. For instance, **E** is sign-regular. Therefore, $\mathbf{A} = \mathbf{E}^T \mathbf{L}^{-1} \mathbf{E}$ is also sign-regular and consequently $\mathbf{A}^{-1}$ is totally non-negative. Furthermore, since **A** exists and the off-diagonal terms of $\mathbf{A}^{-1}$ are positive, $\mathbf{A}^{-1}$ is in fact oscillatory.

2. Properties of oscillatory matrices

We shall continue to touch very selectively upon those results of the theory of oscillatory matrices which are useful for inverse eigenvalue problems.

Since all the elements of an oscillatory matrix are non-negative a matrix **Q** is decomposable (reducible) if by suitable and simultaneous permutations of rows and columns it can be transformed into

$$\mathbf{P}\,\mathbf{Q}\,\mathbf{P}^T = \begin{vmatrix} \mathbf{B} & \mathbf{O} \\ \mathbf{C} & \mathbf{D} \end{vmatrix}$$

where **B** and **D** are square matrices and **O** is a rectangular matrix of zeros. Now, if **Q** is a decomposable, oscillatory matrix, then it is easy to check that $(\mathbf{P}\,\mathbf{Q}\,\mathbf{P}^T)^K$ is also decomposable and that it has the same structure as $\mathbf{P}\,\mathbf{Q}\,\mathbf{P}^T$. But since permutations are unitary transformations, it follows that

$$(\mathbf{P}\,\mathbf{Q}\,\mathbf{P}^T)^K = \mathbf{P}\,\mathbf{Q}^K\,\mathbf{P}^T.$$

We have thus reached a contradiction since $\mathbf{Q}^K$ is totally positive and hence not decomposable. Therefore, an oscillatory matrix is indecomposable (or irreducible).

The eigenvalues and eigenvectors of non-negative indecomposable matrices have an important property first discovered by Frobenius. Namely, such matrices have a unique, positive largest eigenvalue. Furthermore, the eigenvector associated with this eigenvalue has non-zero components of like sign.

Oscillatory matrices share these properties. In addition, because they are not only non-negative but totally non-negative they have further properties. In order to see how these properties come about, I must remind you of what is a <u>compound matrix</u>.

Let $\mathbf{A} = ||a_{ij}||$ be a generic matrix. Form all possible minors of order p, viz.

$$A \begin{pmatrix} i_1 & i_2 & \cdots & i_p \\ j_1 & j_2 & \cdots & j_p \end{pmatrix}$$

There are Γ^2 such minors where $\Gamma = \binom{N}{p}$ is the number of combinations of N objects taken p at a time. By associating in a lexicographical order indices $\alpha_1, \alpha_2, \ldots, \alpha_\Gamma$ with each of the p-tuples $(i_1 \ldots i_p)$, $1 \leq i_1 < \ldots < i_p \leq N$, we construct a $\Gamma \times \Gamma$ matrix called the p-th compound of $\mathbf{A}$ and denoted by $\mathbf{A}^{(p)}$. Thus, $\mathbf{A}^{(1)}$ is identical to $\mathbf{A}$ whereas $\mathbf{A}^{(\Gamma)}$ is a 1x1 matrix consisting of the single element $|A|$. It is possible to show that if

$$\mathbf{C} = \mathbf{A}\,\mathbf{B}$$

then

$$\mathbf{C}^{(p)} = \mathbf{A}^{(p)}\;\mathbf{B}^{(p)}$$

and if

$$\mathbf{B} = \mathbf{A}^{-1}$$

then

$$\mathbf{B}^{(p)} = [\,\mathbf{A}^{(p)}\,]^{-1}.$$

We are now in a position to understand the properties of oscillatory matrices. As previously mentioned, if $\mathbf{A}$ is oscillatory, then it is non-negative and undecomposable. As such it has a unique positive largest eigenvalue, say r_1. The 2nd compound $\mathbf{A}^{(2)}$, is also a non-negative undecomposable matrix. Hence Frobenius theorem implies the existence of a unique positive largest eigenvalue which is $r_1 r_2$, where r_2 is the second largest eigenvalue of $\mathbf{A}$. As a result, the oscillatory matrix $\mathbf{A}$ has N distinct, positive eigenvalues. In addition, the eigenvector associated with the nth eigenvalue has n-1 variations of sign. The various concepts mentioned above have a counterpart in the theory of integral equations.

3. Oscillatory kernels

Kellogg (1916, 1918) was the first to show that if the real symmetric kernel $K(x,y)$ defined on the square $(0,L)\times(0,L)$ is such that the determinants

$$K\begin{pmatrix} x_1 & x_2 & \cdots & x_n \\ y_1 & y_2 & \cdots & y_n \end{pmatrix} = \begin{vmatrix} K(x_1,y_1) & K(x_1,y_2) & \cdots & K(x_1,y_n) \\ K(x_2,y_1) & K(x_2,y_2) & \cdots & K(x_2,y_n) \\ K(x_n,y_1) & K(x_n,y_2) & \cdots & K(x_n,y_n) \end{vmatrix}$$

are non-negative and

$$K\begin{pmatrix} x_1 & x_2 & \cdots & x_n \\ x_1 & x_2 & \cdots & x_n \end{pmatrix}$$

are positive for all partitions

$$0 < \begin{matrix} x_1<x_2< \ldots <x_n \\ y_1<y_2< \ldots <y_n \end{matrix} < L,$$

and for $n=1,2,3,\ldots$, then the eigensolutions of the integral equation

$$\omega^{-2}\,u(x) = \int_0^L K(x,y)\,u(y)\,dy$$

have the following remarkable properties:
(i) all the eigenvalues are simple and positive,
(ii) the eigenfunction associated with the nth eigenvalue has n-1 sign reversals.

The eigenfunctions have further properties such as the interlacing of the zeros of two consecutive eigenfunctions which we do not bother to mention. Such eigenfunctions form a Chebychev system (Karlin 1966).

These results have been extended to asymmetric kernels by Gantmakher (1936) as well as to integral equations with weight $\rho(x)$ by Krein. In his seminal papers, Kellogg raises the problem of characterizing those differential operators whose Green functions are oscillating kernels. This question was answered by Krein (1936,1939). He showed that the eigenvalue problem associated with the operator

$$\mathcal{L} = \alpha_1 d/dx\ \alpha_2 d/dx\ \ldots\ d/dx(\alpha_n\ldots\alpha_1)^{-2} d/dx\ \ldots\ d/dx\ \alpha_2 d/dx\ \alpha_1,$$

i.e.

$$(-1)^n\,\mathcal{L}u = \omega^2 \rho u, \qquad 0 < x < L,$$

and a wide class of boundary conditions at $x=0$ and $x=L$ can be transformed into an integral equation with an oscillatory kernel provided that all the α's are positive.

I want to dwell on this result and outline the proof since it is not given in Krein's paper. The first few steps require standard results from the theory of ordinary differential equations which can be found in Ince (1956, p.120).

Consider a linear differential equation of order m. It has m linearly independent solutions $y_1, \ldots, y_m$. The differential equation, whose solutions are $y_1, \ldots, y_m$, can therefore be written thus:

$$W(y\ y_1 \ldots y_m) = \begin{vmatrix} y & y_1 & \cdots & y_m \\ y' & y_1' & \cdots & y_m' \\ \vdots & \vdots & & \vdots \\ y^{(m)} & y_1^{(m)} & \cdots & y_m^{(m)} \end{vmatrix} = 0.$$

or, dividing by the wronskian $W(y_1 \ldots y_m)$ in order for the coefficient of $y^{(m)}$ to be one, we can also write

$$D_m \,/\, d_m = 0$$

where

$$D_m = W(y\ y_1 \ldots y_m),$$

$$d_m = W(y_1 \ldots y_m).$$

The next step consists in putting in evidence the m derivatives. This is accomplished by means of the repeated use of the following identity:

$$D_r\, d_{r-1} = D_{r-1}\, d_r' - D_{r-1}'\, d_r .$$

In order to prove this identity, we form a two-block matrix whose determinant is obviously equal to $D_r\, d_{r-1}$, namely

$$\begin{vmatrix} y & y_1 & \cdots & y_{r-1} & y_r & 0 & \cdots & 0 \\ \vdots & \vdots & & \vdots & \vdots & \vdots & \cdots & \vdots \\ y^{(r-1)} & y_1^{(r-1)} & \cdots & y_{r-1}^{(r-1)} & y_r^{(r-1)} & 0 & \cdots & 0 \\ y^{(r)} & y_1^{(r)} & \cdots & y_{r-1}^{(r)} & y_r^{(r)} & 0 & \cdots & 0 \\ 0 & 0 & \cdots & 0 & 0 & y_1 & \cdots & y_{r-1} \\ \vdots & \vdots & & \vdots & \vdots & \vdots & & \vdots \\ 0 & 0 & & 0 & 0 & y_1^{(r-2)} & \cdots & y_{r-1}^{(r-2)} \end{vmatrix} = D_r \, d_{r-1}$$

If we evaluate this determinant via Laplace's formula with minors of order $r+1$ and $r-1$ formed by using the first $r+1$ and last $r-1$ columns respectively, it is clear that we can replace the zeros in the $r+1$ st column by any other numbers. We shall do so. We shall also add the jth column to the $r+j$ th one for $j=2,\ldots,r$. As a result, we get

$$\begin{vmatrix} y & y_1 & \cdots & y_{r-1} & y_r & y_1 & \cdots & y_{r-1} \\ \vdots & \vdots & & \vdots & \vdots & \vdots & \cdots & \vdots \\ y^{(r-1)} & y_1^{(r-1)} & \cdots & y_{r-1}^{(r-1)} & y_r^{(r-1)} & y_1^{(r-1)} & \cdots & y_{r-1}^{(r-1)} \\ y^{(r)} & y_1^{(r)} & \cdots & y_{r-1}^{(r)} & y_r^{(r)} & y_1^{(r)} & \cdots & y_{r-1}^{(r)} \\ 0 & 0 & \cdots & 0 & y_r & y_1 & \cdots & y_{r-1} \\ \vdots & \vdots & & \vdots & \vdots & \vdots & & \vdots \\ 0 & 0 & & 0 & y_r^{(r-2)} & y_1^{(r-2)} & \cdots & y_{r-1}^{(r-2)} \end{vmatrix} = D_r \, d_{r-1}$$

Using once again Laplace's formula for the evaluation of the determinant, but with minors of order r now formed by means of the first and last r columns, we get the desired identity.

By repeated use of the identity, we conclude that the most general linear ordinary differential equation of order m can be written thus:

$$L_m(y) = \frac{d_m}{d_{m-1}} \frac{d}{dx} \frac{d_{m-1}^2}{d_m d_{m-2}} \frac{d}{dx} \cdots \frac{d}{dx} \frac{d_1^2}{d_2 d_0} \frac{d}{dx} \frac{d_0}{d_1} y.$$

where $d_0 = 1$. The above expression is greatly simplified if we recall that the m linearly independent solutions can always be written as follows:

$$y_1 = \alpha_1^{-1},$$

$$y_2 = \alpha_1^{-1} \int^x \alpha_2^{-1}\, dt_1,$$

$$\ldots\ldots\ldots\ldots\ldots$$

$$y_m = \alpha_1^{-1} \int^x \alpha_2^{-1}\, dt_1 \int^{t_1} \ldots \int^{t_{m-2}} \alpha_m^{-1}\, dt_{m-1}.$$

Making use of these expressions, we can state that the most general linear differential operator of order m has the form

$$L_m(y) = (\alpha_1\alpha_2\ldots\alpha_m)^{-1} \frac{d}{dx}\, \alpha_m \frac{d}{dx} \ \ldots\ \frac{d}{dx}\, \alpha_2 \frac{d}{dx}\, \alpha_1 y.$$

At this point, we restrict our attention to self-adjoint operators. Their order must be even, say 2n, and the above structure becomes

$$L_{2n}(y) = \alpha_1 \frac{d}{dx}\, \alpha_2 \frac{d}{dx} \ldots \alpha_n \frac{d}{dx}\, P^{-2} \frac{d}{dx}\, \alpha_n \ldots \frac{d}{dx}\, \alpha_2 \frac{d}{dx}\, \alpha_1 y.$$

The factor P is the product of the various coefficients

$$P = \alpha_1\alpha_2 \ldots \alpha_n.$$

Of course, we could relax our requirement that the coefficient of the highest derivative have absolute value 1 and treat P as a function.

Now, let us consider the eigenvalue problem

$$(-1)^n L_{2n}(y) = \omega^2 P y \qquad x \in (0,L)$$

$$y = y^{(1)} = y^{(2)} = \ldots = y^{(n-1)} = 0 \quad \text{at } x=0,L.$$

Given that

$$\alpha_i > 0 \quad \text{for } i=1,2,\ldots,n$$

a proof that the corresponding Green function is oscillatory can be found in Karlin (1968, p.545). We want to consider the converse, i.e show that given that

$$G\begin{pmatrix} x_1 & x_2 & \ldots & x_m \\ s_1 & s_2 & \ldots & s_m \end{pmatrix} \geq 0,$$

with the inequality holding for $x_i = s_i$, i.e. that $G(x,s)$ is a Kellogg kernel, then

$$\alpha_i > 0 \quad \text{for } i=1,2,\dots,n.$$

The proof is in two steps and relies heavily on the fact that $G(x,s)$ is not just any Kellogg kernel but in fact a Green function. Let us therefore review some of the properties of Green functions in general and $G(x,s)$ in particular.

The Green function $G(x,s)$ has the following representation:

$$G(x,s) = \begin{cases} \sum_{i=1}^{n} W_i(s)\,\eta_i(x) & x \geq s, \\ -\sum_{i=n+1}^{2n} W_i(s)\,\eta_i(x) & x \leq s. \end{cases}$$

where $\{\eta_i\}_1^{2n}$ are linearly independent solutions of $L_{2n}(y)=0$. The proof follows from the fact that at $x=s$, we have

$$\begin{vmatrix} \eta_1(s) & \dots & \eta_{2n}(s) \\ \cdot & & \cdot \\ \cdot & & \cdot \\ \cdot & & \cdot \\ \eta_1^{(2n-1)}(s) & \dots & \eta_{2n}^{(2n-1)}(s) \end{vmatrix} \begin{vmatrix} W_1(s) \\ \cdot \\ \cdot \\ \cdot \\ W_{2n}(s) \end{vmatrix} = \begin{vmatrix} 0 \\ \cdot \\ \cdot \\ \cdot \\ 1 \end{vmatrix}$$

We next prove a weaker version of a theorem of Krein, viz.

<u>Theorem</u>: If $G(x,s)$ is (i) totally non-negative

(ii) a Green function

then $G(x,s) > 0$ for $0 < x,s < L$.

Proof: Let us assume that there exists an x_0 and an s_0 such that $G(x_0,s_0) = 0$.

Consider the determinant

$$\begin{vmatrix} G(x_0,s_0) & G(x_0,s) \\ G(x,s_0) & G(x,s) \end{vmatrix}$$

where $x_0 < x$ and $s_0 < s$. On account of the total non-negativity of G and of the properties of x_0 and s_0 we conclude that

$$-G(x_0,s)\,G(x,s_0) \geq 0,$$

i.e either $G(x_0,s)=0$ or $G(x,s_0)=0$. The reasoning in the case in which

$$G(x,s_0)=0 \qquad \text{for } x_0 < x < L$$

is typical of both cases. So let us assume that this is the case. Let us also assume that $s_0 < x_0$ since otherwise we immediately reach a contradiction, namely that the 2n-1st derivative of G is not discontinuous at $x=s_0$. From the representation of G, it follows that

$$\sum_{i=1}^{n} W_i(s_0)\,\eta_i(x) = 0 \qquad \text{for } x_0 < x < L$$

which in view of the independence of the $\{\eta_i\}_1^n$ implies that

$$W_i(s_0) = 0 \qquad \text{for } i=1,\dots,n.$$

As a result

$$G(x,s_0) = 0 \qquad \text{for } s_0 \leqq x < L.$$

Returning once again to the representation of G, we see that the above result implies that $\{\eta_i\}_{n+1}^{2n}$ are not independent which is not possible.

Krein has in fact proved a more general theorem about the positivity of G and its iterates in various subrectangular regions of the square $(0,L)\times(0,L)$. One important consequence of this theorem is that all totally non-negative Green functions are oscillatory functions. In other words: if (i) $G(x,s)$ is such that $L_m G = \delta(x-s)$ and (ii)

$$G\begin{pmatrix} x_1 & \dots & x_n \\ s_1 & \dots & s_n \end{pmatrix} \geqq 0 \quad \text{for} \quad 0 < \begin{matrix} x_1 < \dots < x_n \\ s_1 < \dots < s_n \end{matrix} < L$$

then $$G\begin{pmatrix} x_1 & \dots & x_n \\ x_1 & \dots & x_n \end{pmatrix} > 0.$$

Said in yet another way, totally non-negative Green functions are Kellogg kernels.

We are finally in a position to answer the question about which linear differential operators give rise to oscillatory Green functions.

<u>Theorem (Krein 1936)</u>: If G is (i) a Green function such that

$$L_{2n}G = \alpha_1 \frac{d}{dx} \alpha_2 \frac{d}{dx} \quad \dots \quad \frac{d}{dx} \alpha_2 \frac{d}{dx} \alpha_1 G = \delta(x-s)$$

(ii) oscillatory

(iii) symmetric

then:

$$\alpha_i > 0 \text{ for } i=1,\dots,n.$$

Proof: For $x \geqq s$ we can deduce from our former representation after taking into account the boundary conditions that

$$G = \frac{s^n}{n!} \sum_{i=1}^{n} W_i^{(n)}(ts)\ \eta_i(x) , \qquad \text{where } 0 \leq t \leq 1.$$

From the positivity of G, we conclude that as $s \to 0$

$$\alpha_1^{-1} = \sum_{i=1}^{n} W_i^{(n)}(0)\ \eta_i(x) \quad > 0.$$

Similarly, the oscillatory nature of G implies that

$$\begin{vmatrix} G(x.s) & \partial G(x,s)/\partial s \\ \partial G(x,s)/\partial x & \partial^2 G(x,s)/\partial x \partial s \end{vmatrix} > 0$$

or equivalently

$$\begin{vmatrix} \sum_{i=1}^{n} W_i^{(n)}(ts)\ \eta_i(x) & \sum_{i=1}^{n} W_i^{(n+1)}(ts)\ \eta_i(x) \\ \sum_{i=1}^{n} W_i^{(n)}(ts)\ \eta_i^{(1)}(x) & \sum_{i=1}^{n} W_i^{(n+1)}(ts)\ \eta_i^{(1)}(x) \end{vmatrix} > 0$$

where t is a number between 0 and 1. By letting s tend to zero and defining a second solution y_2 thus

$$y_2 = y_1 \int \alpha_2^{-1}\, dx = \sum_{i=1}^{n} W_i^{(n+1)}(0)\ \eta_i(x)$$

it follows that $\alpha_2 > 0$. Repeating this procedure we arrive at the desired result.

V. INVERSE PROBLEM FOR THE VIBRATING BEAM

1. Introduction

The inverse problem for the vibrating beam provides a test of our understanding of the correspondiong problem for the string. Indeed, this is where we sort out which results can be generalized and which methods can be extended. In addition, the problem for the beam gives rise to certain questions which have no counterpart in the simpler theory of the inverse string problem.

An investigation of the problem for the most general boundary conditions is unnecessarily complicated. For this reason, we shall confine our attention to a single vibrating configuration, namely that in which the left end is free and the right one is clamped.

Thus, the central eigenvalue problem is:

$$(r(x)u_n''(x))'' = \omega_n^2 \rho(x)u_n(x)$$

$$ru_n'' = (ru_n'')' = 0, \text{ at } x = 0$$

$$u_n = u_n' = 0, \text{ at } x = L$$

Note that the differential operator satisfies the conditions of Krein's theorem. Hence, the Green function will be an oscillatory kernel.

To allow for the possibility that the flexural rigidity $r(x)$ is not twice differentiable, we rewrite the basic eigenvalue problem thus:

$$u_n'(x) = \theta_n(x)$$

$$\theta_n'(x) = r^{-1}(x)\tau_n(x)$$

$$\tau_n'(x) = -\phi_n(x)$$

$$\phi_n'(x) = -\omega_n^2 \rho(x)u_n(x)$$

In these equations $u_n(x)$ and $\theta_n(x)$ are the displacement and slope of the center line of the nth normal mode, $\tau_n(x)$ and $\phi_n(x)$ the moment and shearing force about this line. Throughtout our discussion, we shall assume that the eigenfunctions are normalized thus:

$$\int_0^L \rho(x)\, u_m(x)\, u_n(x)\, dx = \delta_{mn}$$

We shall also adopt a further normalization regarding the signs of the eigenfunctions, namely

$$u_n(0) > 0, \qquad n=1,2, \ldots$$

With this sign convention, the signs of $\Theta_n(0)$ are automatically determined. Indeed, by repeated integration of the basic eigenvalue equation it is easy to show that

$$\Theta_n(x) = \Theta_n(0) + w_n^2 \int_0^x r^{-1}(t)\, dt \int_0^t (t-t')\, \rho(t')\, u_n(t')\, dt'$$

By selecting x to coincide with the first zero of $u_n(x)$, the above formula implies that

$$\Theta_n(0) < 0, \qquad n=1,2, \ldots$$

We know that asymptotic trends will be required. For future reference we recall those of w_n, u_n and Θ_n:

$$w_n = O(n - 1/2)^2 ,$$

$$u_n(0) = O(1) ,$$

$$\Theta_n(0) = O(n - 1/2) .$$

We have already alluded to the properties of the Green function. We record here its explicit representation:

$$G(x,\xi) = \begin{cases} \int_\xi^L r^{-1}(t)\, (t - x)(t - \xi)\, dt, & x < \xi \\ \\ \int_x^L r^{-1}(t)\, (t - x)(t - \xi)\, dt, & x > \xi \end{cases}$$

as well as its eigenfunction representation:

$$G(x,\xi) = \sum_1^\infty w_n^{-2}\, u_n(x)\, u_n(\xi)$$

From these expressions for the Green function, we deduce that

$$\sum_{n=1}^{\infty} \omega_n^{-2}\, u_n^2(x) = \int_x^L r^{-1}(t)\,(t-x)^2\, dt$$

$$\sum_{n=1}^{\infty} \omega_n^{-2}\, u_n(x)\, \theta_n(x) = -\int_x^L r^{-1}(t)\,(t-x)\, dt$$

$$\sum_{n=1}^{\infty} \omega_n^{-2}\, \theta_n^2(x) = \int_x^L r^{-1}(t)\, dt$$

as well as

$$\sum_{n=1}^{\infty} \omega_n^{-2} = \int_0^L \rho(x)\, dx \int_x^L r^{-1}(t)\,(t-x)^2\, dt$$

We have made use of similar relations in our solution of the string problem. Not surprisingly, these relations will also play a crucial role in the solution of the beam problem.

2. Uniqueness : equivalent data sets

The inverse problem consists in determining the density $\rho(x)$ and the flexural rigidity $r(x)$ from spectral data. Just as for the string, the impulse response points to the necessary data.

If we impulsively strike the free left end of the beam, then the impulse response consists of the temporal records of both the displacement at $x=0$ and the slope of the mid-line. Using a caret to denote Laplace transforms, we can show by means of considerations analogous to those for the string (Barcilon, 1982) that

$$\hat{y}(0,s) = F_2\, \frac{\prod_n^{\infty} (1 + s^2/\mu_n^2)}{\prod_n^{\infty} (1 + s^2/\omega_n^2)},$$

$$\hat{\theta}(0,s) = F_1\, \frac{\prod_n^{\infty} (1 + s^2/\nu_n^2)}{\prod_n^{\infty} (1 + s^2/\omega_n^2)},$$

where $\{\mu_n\}$ and $\{\nu_n\}$ are the eigenvalues of the same beam but with boundary conditions $u=T=0$ and $u=\phi=0$ respectively at $x=0$; finally,

$$F_m = \int_0^L t^m dt \,/\, r(t).$$

One can take the three-spectra data, consisting of $\{\mu_n\}$, $\{\nu_n\}$, $\{\omega_n\}$, F_1, F_2, as the starting point of the inversion procedure. However, as we shall see, such a data set must satisfy certain conditions for a physical solution to exist. The search for these conditions is rather involved for the three-spectra data. For this reason, the one-spectrum/end-point data set is preferable.

To see the relationship between these two data sets, we write the impulse response thus:

$$\hat{y}(0,s) = \sum_n^{\infty} \frac{u_n^2(0)}{s^2 + \omega_n^2},$$

$$\hat{\theta}(0,s) = \sum_n^{\infty} \frac{u_n(0)\theta_n(0)}{s^2 + \omega_n^2}.$$

Clearly, the one-spectrum/end-point data, consisting of $\{u_n(0), \theta_n(0), \omega_n\}$, are equivalent to the three-spectra data. They will prove easier to deal with when it comes to check whether or not they are bona fide data. Note that the analog of the formula for the length of the string are:

$$F_2 = \sum_{n=1}^{\infty} \frac{u_n^2(0)}{\omega_n^2},$$

$$F_1 = - \sum_{n=1}^{\infty} \frac{u_n(0)\theta_n(0)}{\omega_n^2}.$$

Finally, note that the sign of $u_n(0)\theta_n(0)$ has already played a role.

3. Discretization

Once again, we shall rely upon a discretization to construct an approximate solution. Just as in the case of the string, the discretization must take cognizance of the amount of data in a truncated set. In particular, we must insure that our discrete beam has as many unknown elements as we have data. To obtain such a beam, we replace the functions $\rho(x)$ and $1/r(x)$ by a finite sum of delta functions as follows:

$$\rho^{(N)}(x) = \sum_{n=0}^{N} m_i \delta(x - x_i)$$

$$1/r^{(N)}(x) = \sum_{n=1}^{N} f_i \delta(x - x_i)$$

where

$$x_0 = 0 +$$

Substituting the above expressions for $\rho(x)$ and $1/r(x)$ in the governing equations, we obtain the discrete equations considered by Gladwell (1984)

$$\theta_i = (u_{i+1} - u_i)/l_i \quad ,$$

$$f_i \tau_i = \theta_i - \theta_{i-1} \quad ,$$

$$\phi_i = -(\tau_i - \tau_{i-1})/l_{i-1} \quad ,$$

$$-\omega^2 m_i u_i = \phi_{i+1} - \phi_i \quad ,$$

where

$$u_{n,i} = u_n(x_i + 0) \quad ,$$

$$\theta_{n,i} = \theta_n(x_i + 0) \quad ,$$

$$\tau_{n,i} = \tau_n(x_i - 0) \quad ,$$

$$\phi_{n,i} = \phi_n(x_i - 0) \quad .$$

We continue to use the first subscript to refer to the nodal number and the second to indicate position, i.e. we continue to use a notation reminiscent of the continuum case. We shall suppress the nodal index whenever possible. Finally, we let

$$l_i = x_{i+1} - x_i \quad .$$

The boundary conditions are :

$$\tau_0 = \phi_0 = 0 \quad ,$$

$$u_{N+1} = \theta_N = 0 \quad .$$

Note that they imply that $u_N = 0$, i.e. that the N-th mass does not move. Hence the above discrete eigenvalue problem has N eigenvalues.

If we define four column vectors of length N thus:

$$\mathbf{u}^T = (u_0, \ldots, u_{N-1}) \quad ,$$

$$\boldsymbol{\theta}^T = (\theta_0, \ldots, \theta_{N-1}) \quad ,$$

$$\boldsymbol{\tau}^T = (\tau_1, \ldots, \tau_N) \quad ,$$

$$\boldsymbol{\phi}^T = (\phi_1, \ldots, \phi_N) \quad ,$$

we can write the discrete equations in matrix notation:

$$\boldsymbol{\theta} = -\mathbf{L}^{-1}\mathbf{E}\,\mathbf{u}$$

$$\boldsymbol{\tau} = -\mathbf{R}\,\mathbf{E}\,\boldsymbol{\theta}$$

$$\boldsymbol{\phi} = -\mathbf{L}^{-1}\,\mathbf{E}^T\,\boldsymbol{\tau}$$

$$\omega^2\,\mathbf{M}\,\mathbf{u} = -\mathbf{E}^T\,\boldsymbol{\phi}$$

where **R** is diagonal matrix with entries $(f_1^{-1}, \ldots, f_N^{-1})$ and **L**, **M**, and **E** are all matrices previously encountered.

Given $\{u_{n,0}\}_{n=1}^N$, $\{\theta_{n,0}\}_{n=1}^N$ and $\{\omega_n\}_{n=1}^N$, the discrete inverse problem consists in determining $\{m_i\}_0^{N-1}$, $\{f_i\}_1^N$ and $\{l_i\}_0^{N-1}$ since m_N cannot be determined in this vibrating configuration.

We follow the approach given by Gladwell (1984) to solve for the discrete beam. The first step in this approach is to note that the completeness of the eigenvectors $\{\mathbf{u}_n\}$ and $\{\tau_n\}$ implies that

$$\sum_{n=1}^N u_{n,i}u_{n,j} = m_i^{-1}\,\delta_{ij}, \qquad i=0, \ldots, N-1$$

$$\sum_{n=1}^N \omega_n^{-2}\,\tau_{n,i}\tau_{n,j} = f_i^{-1}\,\delta_{ij}, \qquad i=1, \ldots, N$$

These relations are used to compute the point masses and concentrated flexural rigidities. The computation of the length intervals proceeds from the geometric definition of the slope together with the completeness condition. Indeed, after multiplication by $u_{n,i}$ and summation over n, we see that

$$\sum_{n=1}^N \theta_{n,i}u_{n,i} = -\,l_i^{-1}\sum_{n=1}^N u_{n,i}u_{n,i}, \qquad i=0, \ldots, N-1$$

The procedure is thus clear. Given: $u_{n,i}$, $\theta_{n,i}$, $\phi_{n,i}$, $\tau_{n,i}$, we can compute m_i, l_i, f_i as well as $u_{n,i+1}$, $\theta_{n,i+1}$, $\phi_{n,i+1}$, $\tau_{n,i+1}$ as follows:

$$m_i^{-1} = \sum_{n=1}^N u_{n,i}^2,$$

$$l_i^{-1} = - m_i \sum_{n=1}^{N} \theta_{n,i} u_{n,i},$$

$$u_{n,i+1} = u_{n,i} + l_i \theta_{n,i},$$

$$\phi_{n,i+1} = \phi_{n,i} - \omega_n^2 m_i u_{n,i},$$

$$\tau_{n,i+1} = \tau_{n,i} - l_i \phi_{n,i+1},$$

$$f_{i+1}^{-1} = \sum_{n=1}^{N} \omega_n^{-2} \tau_{n,i+1}^2,$$

$$\theta_{n,i+1} = \theta_{n,i} + f_{i+1} \tau_{n,i+1}.$$

The above recurrent scheme holds for $i=0, \ldots, N-1$. Thus, the procedure used for the string has been generalized to the beam.

Two problems remain: first, we must examine whether the intervals $\{l_i\}$ thus constructed are positive; second, we must consider the limit $N\to\infty$.

We shall postpone the question of the positivity of the intervals and sketch the limiting process.

4. The continuum limit

We set the stage for the limiting process $N\to\infty$, by deriving the discrete analogs of Green function traces. To that effect we write the matrix eigenvalue problem thus:

$$\mathbf{A}\,\mathbf{u}_n = \omega_n^2\, \mathbf{M}\,\mathbf{u}_n$$

where

$$\mathbf{A} = \mathbf{E}^T \mathbf{L}^{-1} \mathbf{E}^T \mathbf{R}\,\mathbf{E}\,\mathbf{L}^{-1} \mathbf{E}$$

Since $\mathbf{A}^{-1}$ and not $\mathbf{A}$ is an oscillatory matrix, we rewite the eigenvalue problem as

$$\mathbf{U}\,\mathbf{\Omega}^{-1} = \mathbf{A}^{-1} \mathbf{M}\,\mathbf{U}$$

where Ω is a diagonal matrix with entries ($\omega_1^2, \ldots, \omega_N^2$). On account of the completeness relation, we deduce that:

$$\mathbf{A}^{-1} = \mathbf{U}\,\mathbf{\Omega}^{-1}\,\mathbf{U}^T.$$

By evaluating both sides explicitly (Barcilon, 1986) we find one of the desired relations

$$\sum_{n=1}^{N} \omega_n^{-2} u_{n,i-1}^2 = \sum_{k=i}^{N} (x_k - x_{i-1})^2 f_k , \qquad i=1,\ldots,N.$$

A second, more useful relation, is found by pre- and post-multiplying the formula for $\mathbf{A}^{-1}$ by $\mathbf{L}^{-1}\mathbf{E}$ and $\mathbf{E}^T\mathbf{L}^{-1}$ respectively. This yields

$$\mathbf{\Theta}\,\mathbf{\Omega}^{-1}\,\mathbf{\Theta}^T = \mathbf{E}^{-1}\,\mathbf{R}^{-1}\,\mathbf{E}^{-T}.$$

Carrying out once again the calculation of both sides explicitly with the help of

$$\mathbf{E}^{-1} = \begin{vmatrix} 1 & 1 & 1 & \ldots & 1 \\ 0 & 1 & 1 & \ldots & 1 \\ 0 & 0 & 1 & \ldots & 1 \\ \vdots & \vdots & \vdots & & \vdots \\ 0 & 0 & 0 & & 1 \end{vmatrix}$$

we find that

$$\sum_{n=1}^{N} \omega_n^{-2} \theta_{n,i-1}^2 = \sum_{k=i}^{N} f_k^{(N)}, \qquad i=1,2,\ldots,N.$$

Finally, the last relation is obtained by considering the trace of $\mathbf{A}^{-1}\mathbf{M}$, namely

$$\sum_{n=1}^{N} \omega_n^{-2} = \sum_{i=1}^{N} m_{i-1} \{ \sum_{k=i}^{N} (x_k - x_{i-1})^2 f_k \}.$$

After these preliminaries, we reinstate the superscripts which indicate the order of the truncation and define

$$S^{(N)}(x) = \begin{cases} 0 & \text{for } x \geq x_N^{(N)} \\ \sum_{k=i}^{N} f_k^{(N)} & \text{for } x_{i-1}^{(N)} \leq x < x_i^{(N)} \end{cases} \qquad i=1,\ldots,N.$$

From the above definition, we can see that the sequence $\{S^{(N)}(x)\}$ is made up of non-negative, non-increasing functions of x. Furthermore, we know that

$$S^{(N)}(x) \leq S^{(N)}(0) = \sum_{n=1}^{N} \omega_n^{-2} \theta_{n,0}^2 .$$

Therefore,

$$S^{(N)}(x) \leq F_0 = \sum_{n=1}^{\infty} \omega_n^{-2} \theta_n^2 (0),$$

i.e. the terms of the sequence $\{S^{(N)}(x)\}$ are uniformly bounded. Hence, by Helly's selection theorem, this sequence has a subsequence which converges to a non-negative, non-increasing function

$$S(x) = \int_x^{\infty} r^{-1} \, dt .$$

The uniqueness results guarantee the uniqueness of this limiting function.Where S(x) is differentiable we can define a rigidity r(x).

Similarly, we define another sequence of functions $\{Q^{(N)}(x)\}$ where

$$Q^{(N)}(x) = \begin{cases} 0 & \text{for } x_N^{(N)} \leq x \\ & i=1,\ldots,N, \\ \sum_{k=i}^{N} m_{k-1}^{(N)} F^{(N)}(x_{k-1}) & \text{for } x_{i-1}^{(N)} \leq x < x_i^{(N)} \end{cases}$$

and

$$P^{(N)}(x) = \begin{cases} 0 & \text{for } x \geq x_N^{(N)} \\ \sum_{k=1}^{N} f_k^{(N)} (x_k^{(N)} - x_{i-1}^{(N)})^2 & \text{for } x_{i-1}^{(N)} \leq x < x_i^{(N)} \end{cases}$$

The functions $Q^{(N)}(x)$ are also uniformly bounded. Indeed

$$Q^{(N)}(x) \leq Q^{(N)}(0) = \sum_{n=1}^{N} \omega_n^{-2} \leq \sum_{n=1}^{\infty} \omega_n^{-2}$$

Using the same arguments as before, we deduce the existence of a non-negative, non-increasing function Q(x) which a weighted integral of the running mass

$$M(x) = \int_0^x \rho(t) \, d(t) .$$

VI. INVERSE PROBLEM FOR THE VIBRATING BEAM: EXISTENCE

1. Existence of solution

We have been using the one spectrum/end-point data set, namely three sequences $\{u_n(0), \theta_n(0), \{\omega_n\}_1^\infty$ as our starting point. Provided that a solution exists, these data guarantee its uniqueness.

The question is: what are the conditions, over and above the asympotic conditions, which these data must satisfy for a solution to the inverse problem to exist. Note that the length L is not part of the data. For this reason, we shall often think of the beam as being infinite in length but with $r^{-1}(x) = 0$ for $x > L$.

We have already remarked that the differential operator for the eigenvalue problem for the vibrating beam satisfies the conditions of Krein's theorem. Therefore, not only is the Green function $G(x,s)$ an oscillatory kernel, but the eigenfunctions form a Chebyshev set of functions, i.e. the determinants

$$\begin{vmatrix} u_1(x_1) & \dots & u_k(x_1) \\ \vdots & & \vdots \\ u_1(x_k) & \dots & u_k(x_k) \end{vmatrix}$$

for $k=1,2,\dots$ and arbitrary partitions $x_1 < \dots < x_k$ never vanish. This property of the Chebyshev set $\{u_n(x)\}$ is closely related to the fact that the zeros of two consecutive functions interlace. Therefore <u>with our normalization</u> of the eigenfunctions

$$u_1(x) > 0$$

$$\begin{vmatrix} u_1(x_1) & u_2(x_1) \\ u_1(x_2) & u_2(x_2) \end{vmatrix} < 0$$

$$\begin{vmatrix} u_1(x_1) & u_2(x_1) & u_3(x_1) \\ u_1(x_2) & u_2(x_2) & u_3(x_2) \\ u_1(x_3) & u_2(x_3) & u_3(x_3) \end{vmatrix} < 0$$

$$\begin{vmatrix} u_1(x_1) & u_2(x_1) & u_3(x_1) & u_4(x_1) \\ u_1(x_2) & u_2(x_2) & u_3(x_2) & u_4(x_2) \\ u_1(x_3) & u_2(x_3) & u_3(x_3) & u_4(x_3) \\ u_1(x_4) & u_2(x_4) & u_3(x_4) & u_4(x_4) \end{vmatrix} > 0$$

etc ... where the sign alternation has a periodicity of 4 and

$$0 < x_1 < \quad \ldots \quad < x_k < L.$$

This alternation can be seen by selecting the partition

$$x_1 < z_1 < z_2 < \ldots < z_{k-1}$$

where z_1, z_2, ... ,z_{k-1} are the zeros of $u_k(x)$.

In addition, as Krein has shown (1939), the derivatives of the eigenfunctions for the beam also form a Chebyshev system, and as a result

$$\theta_1(x_1) < 0,$$

$$\begin{vmatrix} \theta_1(x_1) & \theta_2(x_1) \\ \theta_1(x_2) & \theta_2(x_2) \end{vmatrix} < 0,$$

$$\begin{vmatrix} \theta_1(x_1) & \theta_2(x_1) & \theta_3(x_1) \\ \theta_1(x_2) & \theta_2(x_2) & \theta_3(x_2) \\ \theta_1(x_3) & \theta_2(x_3) & \theta_3(x_3) \end{vmatrix} > 0,$$

$$\begin{vmatrix} \theta_1(x_1) & \theta_2(x_1) & \theta_3(x_1) & \theta_4(x_1) \\ \theta_1(x_2) & \theta_2(x_2) & \theta_3(x_2) & \theta_4(x_2) \\ \theta_1(x_3) & \theta_2(x_3) & \theta_3(x_3) & \theta_4(x_3) \\ \theta_1(x_4) & \theta_2(x_4) & \theta_3(x_4) & \theta_4(x_4) \end{vmatrix} > 0$$

etc...where here too the signs alternate with periodicity 4.

In order to obtain conditions on the data, i.e. on the eigenfunctions at x=0, we let the points x_1, x_2, ... x_k coallesce onto 0. As a result the first group of inequalities become

$$D(1) = \quad u_1(0) > 0$$

$$D(1\ 2\ ) = \begin{vmatrix} u_1(0) & u_2(0) \\ -\Theta_1(0) & -\Theta_2(0) \end{vmatrix} \quad > 0$$

$$D(1\ 2\ 3) = \begin{vmatrix} u_1(0) & u_2(0) & u_3(0) \\ -\Theta_1(0) & -\Theta_2(0) & -\Theta_3(0) \\ \omega_1^2 u_1(0) & \omega_2^2 u_2(0) & \omega_3^2 u_3(0) \end{vmatrix} \quad > 0$$

etc...

Similarly, the inequalities arising from the Θ's imply that

$$\Delta(1) = \quad -\Theta_1(0) > 0$$

$$\Delta(1\ 2) = \begin{vmatrix} -\Theta_1(0) & -\Theta_2(0) \\ \omega_1^2 u_1(0) & \omega_2^2 u_2(0) \end{vmatrix} \quad > 0$$

$$\Delta(1\ 2\ 3) = \begin{vmatrix} -\Theta_1(0) & -\Theta_2(0) & -\Theta_3(0) \\ \omega_1^2 u_1(0) & \omega_2^2 u_2(0) & \omega_3^2 u_3(0) \\ -\omega_1^2 \Theta_1(0) & -\omega_2^2 \Theta_2(0) & -\omega_3^2 \Theta_3(0) \end{vmatrix} \quad > 0$$

etc...

These inequalities are necessary conditions for the existence of a solution to the inverse problem. They were given for the first time by Gladwell (1984).

2. Necessary conditions

Let us assume that

$$D(1\ 2\ .\ .\ .\ i) > 0$$

$$\Delta(1\ 2\ .\ .\ .\ i) > 0$$

for all integers i. These inequalities imply the positivity of several related determinants. If we define

$$D(i_1\ i_2 \ldots i_{2k}) = \begin{vmatrix} u_{i_1} & u_{i_2} & \cdots & u_{i_{2k}} \\ -\Theta_{i_1} & -\Theta_{i_2} & \cdots & -\Theta_{i_{2k}} \\ \vdots & \vdots & & \vdots \\ -\omega_{i_1}^{2k-2}\Theta_{i_1} & -\omega_{i_2}^{2k-2}\Theta_{i_2} & \cdots & -\omega_{i_{2k}}^{2k-2}\Theta_{i_{2k}} \end{vmatrix}$$

and

$$D(i_1\ i_2 \ldots i_{2k+1}) = \begin{vmatrix} u_{i_1} & u_{i_2} & \cdots & u_{i_{2k+1}} \\ -\Theta_{i_1} & \Theta_{i_2} & \cdots & \Theta_{i_{2k+1}} \\ \vdots & \vdots & & \vdots \\ \omega_{i_1}^{2k} u_{i_1} & \omega_{i_2}^{2k} u_{i_2} & \cdots & \omega_{i_{2k+1}}^{2k} u_{i_{2k+1}} \end{vmatrix}$$

as well as

$$\Delta(i_1 i_2 \ldots i_{2k}) = \begin{vmatrix} -\Theta_{i_1} & -\Theta_{i_2} & \cdots & -\Theta_{i_{2k}} \\ \omega_{i_1}^{2} u_{i_1} & \omega_{i_2}^{2} u_{i_2} & \cdots & \omega_{i_{2k}}^{2} u_{i_{2k}} \\ \vdots & \vdots & & \vdots \\ \omega_{i_1}^{2k} u_{i_1} & \omega_{i_2}^{2k} u_{i_2} & \cdots & \omega_{i_{2k}}^{2k} u_{i_{2k}} \end{vmatrix}$$

and

$$\Delta(i_1 i_2 \ldots i_{2k+1}) = \begin{vmatrix} -\Theta_{i_1} & -\Theta_{i_2} & \cdots & -\Theta_{i_{2k+1}} \\ u_{i_1} & u_{i_2} & \cdots & u_{i_{2k+1}} \\ \vdots & \vdots & & \vdots \\ -\omega_{i_1}^{2k}\Theta_{i_1} & -\omega_{i_2}^{2k}\Theta_{i_2} & \cdots & -\omega_{i_{2k+1}}^{2k}\Theta_{i_{2k+1}} \end{vmatrix}$$

then we can prove that all these determinant are positive. We only sketch the proof here since it is contained in Gladwell's paper.

Consider D(1 2 ... 2k) where k is arbitrary but fixed. If we look upon the array of coefficients making up D(1 2 ... 2k) as a matrix, then this matrix is clearly positive definite. Therefore (Beckenbach & Bellman, 1971 p.63)

$$D(1\ 2\ \ldots\ 2k) < u_1(-\omega_2^2\theta_2)\ \ldots\ (-\omega_{2k}^{2k-2}\theta_{2k})$$

Similarly

$$D(1\ 2\ \ldots\ 2k+1) < u_1(-\omega_2^2\theta_2)\ \ldots\ (\omega_{2k+1}^{2k}u_{2k+1})$$

Since $\Delta(1\ 2\ \ldots\ 2k)$ is also positive definite in the same sense as before, we also have

$$\Delta(1\ 2\ \ldots\ 2k) < (-\theta_1)\omega_2^2u_2\ \ldots\ (\omega_{2k}^{2k}u_{2k})$$

and

$$\Delta(1\ 2\ \ldots\ 2k+1) < (-\theta_1)\omega_2^2u_2\ \ldots\ (-\omega_{2k+1}^{2k}\theta_{2k+1})$$

By considering all the values of k sequentially, we deduce that

$$u_n > 0$$

and

$$-\theta_n > 0.$$

thus recovering our normalization. By means of a generalization of the previous inequality for the determinant of positive definite matrices (Beckenbach & Bellman 1971, p.63), we can write

$$D(1\ 2\ \ldots\ 2k) < D(1\ 2\ \ldots\ 2k-3\ 2k-2)\ \omega_{2k-1}^{2k-2}\omega_{2k}^{2k-2}D(2k-1\ 2k)$$

from which we deduce that

$$D(2k-1\ 2k) > 0.$$

Repeating this procedure, we can see that

$$D(2k\ 2k+1) > 0,$$

and more generally that all the D's and Δ's with <u>consecutive</u> indices are positive.

The last step consists in showing that D's and Δ's for arbitrary ranked indices are positive. We first show that $D(i_1\ i_2)$ is positive. If this were not so, then

$$\begin{vmatrix} u_{i_1} & u_{i_2} \\ \theta_{i_1} & \theta_{i_2} \end{vmatrix}$$

would be a positive definite matrix and its determinant would be smaller than $y_{i_1}\theta_{i_2}$ which is negative! Using the same procedure inductively, we can show that

$$D(i_1\ i_2\ \ldots\ i_n) > 0 \qquad n=2,3,\ \ldots$$

Similarly

$$\Delta(i_1\ i_2\ \ldots\ i_n) > 0 \qquad n=2,3,\ \ldots$$

As a result, we can write the necessary conditions for the existence of a solution are equivalent to the total positivity of the infinite matrix

$$\begin{vmatrix} u_1 & u_2 & u_3 & u_4 & \cdots \\ -\theta_1 & -\theta_2 & -\theta_3 & -\theta_4 & \cdots \\ \omega_1^2 u_1 & \omega_2^2 u_2 & \omega_3^2 u_3 & \omega_4^2 u_4 & \cdots \\ -\omega_1^2\theta_1 & -\omega_2^2\theta_2 & -\omega_3^2\theta_3 & -\omega_4^2\theta_4 & \cdots \\ \cdot & \cdot & \cdot & \cdot & \\ \cdot & \cdot & \cdot & \cdot & \\ \cdot & \cdot & \cdot & \cdot & \end{vmatrix}$$

3. Sufficient conditions

Gladwell has shown that the above necessary conditions are also sufficient to guarantee the positivity of the lengths $\{l_i\}$. The following proof is his. The starting point is the matrix equation obtained by post-multiplying the definition of $\mathbf{\Theta}$, viz.

$$\mathbf{\Theta} = -\mathbf{L}^{-1}\mathbf{E}\,\mathbf{U},$$

by $\mathbf{U}^T$ and making use of the completeness relation, namely

$$\mathbf{\Theta}\,\mathbf{U}^T = -\mathbf{L}^{-1}\mathbf{E}\,\mathbf{M}^{-1}.$$

The p-th compound version of this equation reads

$$\mathbf{\Theta}^{(p)}\mathbf{U}^{(p)T} = -\mathbf{L}^{(p)-1}\mathbf{E}^{(p)}\mathbf{M}^{(p)-1}$$

The 1-1 entry in the above equation provides an alternate way to compute the length l_{p-1} .It can be written thus:

$$\sum_{\upsilon=1}^{\mathcal{N}} \theta^{(p)}_{\upsilon,0}\, y^{(p)}_{\upsilon,0} = (-1)^p \{\prod_{i=0}^{p-1} m_i l_i\}^{-1}$$

where

$$\Theta^{(p)}_{\upsilon,0} = \begin{vmatrix} \Theta_{i_1,0} & \cdots & \Theta_{i_p,0} \\ & & \\ \Theta_{i_1,p-1} & \cdots & \Theta_{i_p,p-1} \end{vmatrix} \qquad \upsilon = 1,\ldots,\mathcal{N}$$

As usual

$$\upsilon = (i_1 \ldots i_p)$$

is a lexicographical arrangement of p of the first N integers and

$$\mathcal{N} = N\,!\,/\,p\,!\,(n-p)\,!$$

Finally, $u^{(p)}_{\upsilon,0}$ is given by a similar determinant. By expressing $\Theta_{i,p-1}$, ..., $\Theta_{i,1}$ as well as $u_{i,p-1}$, ..., $u_{i,1}$ in terms of $\Theta_{i,0}$ and $y_{i,0}$ via the discrete version of the governing equations, we can show that $\Theta^{(p)}_{\upsilon,0}$ and $u^{(p)}_{\upsilon,0}$ are related to $\Delta(i_1\ldots i_p)$ and $D(i_1\ldots i_p)$ respectively. These relations contain m_0, l_0, ..., m_{p-2}, l_{p-2} but not l_{p-1}. The positivity of $D(i_1\ldots i_p)$ and $\Delta(i_1\ldots i_p)$, which follows from the necessary conditions, guarantees that $\sum_{\upsilon=1} \Theta^{(p)}_{\upsilon,0}\, u^{(p)}_{\upsilon,0}$ is negative and hence that l_{p-1} is positive.

Thus, for any truncated data set $\{u_n(0),\ \Theta_n(0),\ \omega_n\}_{n=1}^{N}$ which satisfies a truncated set of the necessary conditions, it is possible to construct a discrete beam with positive concentrated masses and rigidities separated by positive intervals.

4. Concluding remarks

I have mentioned at the begining this lecture that there are different equivalent starting points, for solving the inverse problem under consideration, namely the three-spectra approach and the one spectrum/end-point approach.

Given F_1, F_2 and $\{\mu_n,\ \upsilon_n,\ \omega_n\}$ we can compute the one spectrum/end-point data $\{u_n(0),\ \Theta_n(0),\ \omega_n\}$ thus:

$$u^2_n(0) = \omega^2_n F_2 \prod_{k=1}^{\infty} (1-\omega^2_n/\mu^2_k) \;/\; \prod_{k\neq n}^{\infty} (1-\omega^2_n/\omega^2_k)$$

$$\Theta_n(0)u_n(0) = -\,\omega^2_n F_1 \prod_{k=1}^{\infty} (1-\omega^2_n/\upsilon^2_k) \;/\; \prod_{k\neq n}^{\infty} (1-\omega^2_n/\omega^2_k)$$

Hence it is possible to express the necessary and sufficient conditions in terms of the three-spectra data (see Gladwell 1984). However, whereas the kth conditions involve only the kth truncation of the one spectrum/end-point data set, these same kth conditions involve all of the three-spectra data ! Because of this a truncated data set of a bona fide three-spectra data set need not give rise to a real discrete beam. Consequently, one-spetrum/end-point data are preferable than three-spectra data.

Note that the ordering of the eigenvalues, namely

$$\omega_1^2 < \omega_2^2 < \omega_3^2 < \ \ldots \ < \omega_n^2 < \ \ldots$$

is forced by the inequalities and need not be added.

This brings us back to the vibrating string. For this problem, the only condition which the data must satisfy is that the matrix

$$\begin{vmatrix} \omega_1^2 & \omega_2^2 & \omega_3^2 & \ldots \\ \omega_1^4 & \omega_2^4 & \omega_3^4 & \ldots \\ \omega_1^6 & \omega_2^6 & \omega_3^6 & \ldots \\ . & . & . & \\ . & . & . & \\ . & . & . & \end{vmatrix}$$

be totally positive. This condition is automatically satisfied if the eigenvalues are distinct and ordered as we have naturally assumed previously.

In closing, let me summarize the current state of affair. Clearly, the thoery for the string and for the beam is rather complete. Of course, it would be nice to be able to prove uniqueness and existence, as well as obtain a constuction in a more connected manner. But this is purely a question of esthetics. The theory can be extended to the class of inverse eigenvalue problems governed by a differential operator of Krein's form. Eventhough some results are known for general differential operators, the theory is still incomplete. Finally, we still known little about inverse problems for systems of differential operators, i.e about the type of problem arising in the study of the Earth normal modes.

Acknowledgements

I would like to thank Professor G. Talenti for inviting me to present these lectures. Their preparation was supported by the Office of Naval Research under grant N00014-86-K-0035

REFERENCES

Alterman, Z., Jarosch, H. and Pekeris, C. Oscillations of the Earth, Proc. Roy. Soc. Lond., A **252**, (1959) 80-95

Anderssen, R.S. The effect of discontinuities in density and shear velocity on the asymptotic overtone structure of torsional eigenfrequencies of the Earth, Geophys. J. Roy. Astr. Soc., **50**, (1977) 303-309

Barcilon, V. Inverse problem for the vibrating beam in the free-clamped configuration, Phil. Trans. Roy. Soc. Lond., A **304**, (1982) 211-251

———— Explicit solution of the inverse problem for a vibrating string. J. Math. Anal. Appl., **93**, (1983) 222-234

———— Sufficient conditions for the solution of the inverse problem for a vibrating beam, Inverse Prob. (1986) submitted for publication

Backus, G. and Gilbert, F. Numerical applications of a formalism for geophysical inverse problems, Geophys. J. Roy. Astr. Soc., **13**, (1967) 247-276

———— The resolving power of gross earth data, Geophys. J. Roy. Astr. Soc., **16**, (1968) 169-205

Beckenbach, E.F. and Bellman, R. Inequalities, (1971) Spinger-Verlag, NY.

Benioff, H., Press, F. and Smith, S. Excitation of the free oscillations of the Earth by earthquakes, J. Geophys. Res., **66**, (1961) 605-619

Boas, R.P. Entire functions, (1954) Academic Press, NY

Borg, G. Eine Umkehrung der Sturm-liouvilleschen Eigenwertaufgabe, Acta Math., **78**, (1946) 1-96

Gantmakher, F. Sur les noyaux de Kellogg non symetrique, C.R. Acad. Sci. URSS, **1**, (1936) 3-5

Gantmakher, F.P. and Krein, M.G. Sur les matrices completement non negatives et oscillatoires, Compositio Math., **4**, (1937) 445-476

———— Oscillation matrices and kernels and small vibrations of mechanical systems, (1950) State Publishing House for Technical-Theoretical Literature, Moscow-Leningrad. 1961 Translation by US Atomic Energy Commission, Office of Technical Service, Washington, DC.

Gelfand, I.M. and Levitan, B.M. On a simple identity for the eigenvalues of a differential operator of the second order, Dokl. Akad. Nauk SSSR, **88**, (1953) 593-596

Gladwell, G.M.L. The inverse problem for a vibrating beam, Proc. Roy. Soc. Lond., A **393**, (1984) 277-295

Gladwell, G.M.L. and Gbadeyan, J.A. On the inverse problem of the vibrating string or rod, Q.J. Mech. Appl. Math., **38**, (1985) 169-174

Hald, O.H. Discontinuous inverse eigenvalue problems, Comm. Pure Appl. Math., **37**, (1984) 539-577

Hille, E. Lectures on ordinary differential equations, (1968) Addison-Wesley, Reading MA

Hochstadt, H. Asymptotic estimates of the Sturm-Liouville spectrum, Comm. Pure Appl. Math., **14**, (1961) 749-764

Ince, E.L. Ordinary differential equations, (1956) Dover, NY.

Karlin, S. Total positivity, (1968) Stanford University Press

Karlin, S. and Studden, W.J. Tchebycheff systems: with applications in analysis and statistics, (1966) Interscience Publishers, N.Y.

Kellogg, O.D. The oscillation of functions of an orthogonal set, Amer. J. Math., **38**, (1916) 1-5

———————— Orthogonal functions arising from integral equations, Amer. J. Math., **40**, (1918) 145-154

Krein, M.G. Sur les vibrations propres des tiges dont l'une des extremites est encastree et l'autre libre, Comm. Soc. Math. Kharkoff, **12**, (1935) 3-11

———————— Sur les operateurs differentiels oscillatoires, C. R. Acad. Sci. URSS, **4**, (1936) 395-398

———————— Sur les functions de Green non-symetriques oscillatoires des operateurs differentiels ordinaires, C. R. Acad. Sci. URSS, **25**, (1939) 643-646

———————— On inverse problems for an inhomogeneous string, Dokl. Akad. Nauk SSSR, **82**, (1952a) 669-672 [in Russian]

———————— Some new problems in the theory of Sturmian systems, Prikl. Mat. Mekh., **16**, (1952b) 555-568 [in Russian]

Levinson, N. The inverse Sturm-Liouville problem, Mat. Tidssk. B, (1949) 25-30

Marchenko, V.A. Some questions in the theory of one-dimensional linear differential operators of the second order, Trudy Moskov. Mat. Obsc., **1**, (1952) 327-420 [in Russian] translated in Amer. Math. Soc. Trans., **101**, (1973) 1-104

Natanson, I.P. The theory of functions of a real variable, vol I, (1955) F. Ungar Publishing Co., NY.

Poeschel, J. and Trubowitz, E. Lectures on Inverse Spectral Theory, (1984) manuscript.

Richter, F.M. Kelvin and the age of the Earth, J. Geol., to appear

Ness, N.F., Harrison, J.C. and Slichter, L.B. Observations of the free oscillations of the Earth, J. Geophys. Res., **66**, (1961) 621-629

Titchmarch, E.C. Eigenfunction expansions, (1962) Clarendon Press, Oxford

Willis, C. Inverse Sturm-Liouville problems with two discontinuities, Inverse Prob., **1**, (1985) 263-289

REGULARIZATION METHODS FOR LINEAR INVERSE PROBLEMS

M.Bertero
Dipartimento di Fisica dell'Università and
Istituto Nazionale di Fisica Nucleare
I-16146 Genova, Italy

1. INTRODUCTION

In the past two decades, the theory and practice of linear inverse problems have been developed in several domains of applied science: medical diagnostic, atmospheric sounding, radar and sonar target estimation, seismology, radioastronomy, microscopy and so on. Nowadays operational applications are in everyday use, for instance, in X-ray or MR tomography [23], in seismic data processing for geophysical exploration [47] and in radiometric data processing for meteorological forecasts and monitoring [56]. In fact, common procedures for probing a physical sample consist in observing the interaction between the radiation emitted by a known source and the sample or the radiation emitted by the sample itself. One or several detectors measure physical quantities related to the scattered or emitted radiation, such as cross-sections, diffraction or radiation patterns, absorption or reflection coefficients, field fluctuations etc.. The result of the experiment is a real or vector valued function g, depending on one or several variables such as space, time, wavelength of the incident or emitted radiation etc.. The purpose of the experiment is the estimation, through suitable processing of the data function g, of some physical or geometric characteristic of the sample which, in general, can be conceived as a real or vector valued function f, depending on variables whose physical interpretation can be left unspecified. We will call f the object or the object function.

The problem which consists in the determination of the mapping from the set of all possible objects into the set of all possible data is usually called the direct problem. In most cases it implies, as an intermediate step, the solution of a wave equation, the object function f being related to the coefficients of the equation or, eventually, to the boundary conditions. For this reason, the physical nature of the radiation may be irrelevant from the mathematical point of view. Exactly the same mathematical problems may be encountered in completely different physical domains [2,3,8,10,12,13,48].

Then the inverse problem is the determination of the object f from the measured data g. It corresponds to the inversion of the direct mapping, which usually is nonlinear, and therefore the inverse problem implies the solution of a nonlinear functional equation. In most practical cases, however, it is possible and reasonable, on the basis of suitable physical approximations, to linearize the problem. The result is a first kind functional equation

$$A f = g \tag{1.1}$$

where f is the unknown object, g is the data function and A is a known linear operator, obtained by solving the direct problem. If A is an integral operator, then equation (1.1) is a first kind Fredholm integral equation

$$\int_{\Omega} K(x,y)f(y)dy = g(x) \quad , \quad x \in \Omega' \tag{1.2}$$

where $\Omega \subset \mathbb{R}^n$ and $\Omega' \subset \mathbb{R}^m$ are given domains. In some problems, like X-ray tomography, one has to solve an equation similar to (1.2) with a kernel which is not a function but a distribution.

As is well known, first kind Fredholm integral equations are ill-posed in the sense of Hadamard [22]: existence, uniqueness and continuous dependence of the solution on the data usually do not hold true. In fact most, perhaps all linear inverse problems are ill-posed. In the next Section we discuss in detail this point which is basic because we will identify linear inverse and ill-posed problems. This point of view is also supported by the remark that all the classical examples of ill-posed problems, such as the non-standard boundary value problems for partial differential equations [42,9], the analytic and harmonic continuation [11], the Laplace and Fourier transform inversion with limited data [49], provide the mathematical model of several inverse problems.

Finally we point out another important interpretation of equation (1.1). In electrical engineering the process of assigning a function g to a given function f is called a system, with input f and output g. Therefore "system" is a synonym of "operator". Then the solution of equation (1.1) can be viewed as the determination of the input of a known system, given the output g. Since not only electrical devices but also optical and other imaging devices can be described in terms of linear systems [41] , the methods for solving linear inverse problems are of great relevance in many domains of image processing [25].

2. THE ILL-POSEDNESS OF INVERSE PROBLEMS

A precise formulation of problem (1.1) requires a specification of the space X of the objects and of the space Y of the data. In the following we will assume that both X and Y are Hilbert spaces - this assumption is satisfied in most applications - and that $A: X \longrightarrow Y$ is a linear, continuous operator. We will denote by $A^*: Y \longrightarrow X$ the adjoint of A and by N(A) and R(A) respectively the null space and the range of A.

The problem (1.1) is said to be well-posed, in the sense of Hadamard [22], if the following conditions are satisfied:

i) the solution exists for any $g \in Y$;

ii) the solution is unique in X;

iii) the inverse mapping $g \longrightarrow f$ is continuous.

Conditions i) -ii) are equivalent to saying that the operator A has an inverse A^{-1} and that $D(A^{-1}) = Y$. Then, from the linearity and continuity of A and from the open mapping theorem it follows that A^{-1} is also continuous, i.e. in the linear case condition iii) is a consequence of conditions i)-ii).

The requirement of continuous dependence of the solution on the data is a necessary but not sufficient condition for the stability of the solution. In the case of a well-posed problem, relative error propagation from the data to the solution is controlled by the condition number: if δg is a variation of g and δf the corresponding variation of f, then

$$\|\delta f\|_X / \|f\|_X \leqslant \mathrm{cond}(A)\, \|\delta g\|_Y / \|g\|_Y \tag{2.1}$$

where

$$\mathrm{cond}(A) = \|A\|\, \|A^{-1}\| . \tag{2.2}$$

If cond(A) is not too large, the problem (1.1) is said to be <u>well-conditioned</u> and the solution is stable with respect to small variations of the data. Otherwise the problem is said to be <u>ill-conditioned</u>. It is clear that the separation between well-conditioned and ill-conditioned problems is not very sharp and that the concept of well-conditioned problem is more vague than the concept of well-posed problem.

When one of the conditions i)-iii) is not satisfied, the problem (1.1) is <u>ill-posed</u>. It is evident that ill-posedness is a property of the triple $\{A, X, Y\}$: the problem is ill-posed because, for instance, the space Y is too broad. In such a case one could try to transform the ill-posed problem into a well-posed one just by modifying the space Y and its topology. In most practical problem, however, this is a rather useless mathematical excercise because the choice of the appropriate data space and of its topology is not dictated by purely mathematical considerations. We will justify this statement firstly by means of an example and then by some general remarks on the role played by noise - or experimental errors - in an inverse problem.

Consider the Hilbert space $X \subset L^2(\mathbb{R})$ normed by

$$\|f\|_X^2 = \frac{1}{2\pi} \int_{-\infty}^{+\infty} (1 + \xi^{-2})\, |\hat{f}(\xi)|^2\, d\xi \tag{2.3}$$

where $\hat{f}$ denotes the Fourier transform of f, and consider the linear operator A defined on X by

$$(Af)(x) = \int_{-\infty}^{x} f(y)\, dy \quad . \tag{2.4}$$

In such a case the solution of the equation Af=g is just the derivative of g.

If we take $Y = L^2(\mathbb{R})$, then $A: X \longrightarrow Y$ is continuous and has an inverse A^{-1}, but A^{-1} is not continuous because R(A) is not closed in Y. Conditions i) and iii) do not hold true: the solution exists only in the case of differentiable data. On the other hand it may be reasonable to assume that, even in the case of data perturbed by noise, this condition is always satisfied, since it is not very restrictive. The next step, quite natural from the mathematical point of view, is to take as data space a Hilbert space of differentiable functions, for instance $Y = H^1(\mathbb{R})$

$$\| g \|_Y^2 = \frac{1}{2\pi} \int_{-\infty}^{+\infty} (1+\xi^2) \, |\hat{g}(\xi)|^2 \, d\xi \quad . \tag{2.5}$$

Then the operator (2.4) is an isometry, $\| Af \|_Y = \| f \|_X$, and therefore A^{-1} is continuous. The problem is now well-posed but unfortunately the topology of $H^1(\mathbb{R})$ is not appropriate for the description of the perturbation of the data induced by noise.

Assume that the difference between the noisy data g_ε and the exact data g, i.e. the function $w_\varepsilon = g_\varepsilon - g$ which represents the effect of the noise, is a small, strongly oscillating function. An example is provided by

$$w_\varepsilon(x) = \varepsilon \, \phi(x) \cos(x/\varepsilon^2) \tag{2.6}$$

where $\phi \in H^1(\mathbb{R})$ is a given function satisfying, for instance, the condition $|\phi(x)| \leq 1$. If $\varepsilon > 0$ is a small real number, then $w_\varepsilon(x)$ assumes small values and is certainly considered a small perturbation of the data by any physicist or engineer. The L^2-norm of w_ε corresponds to this picture, because it is of the order of ε. On the contrary the H^1-norm of w_ε is of the order of ε^{-1}: in the topology which assures the well-posedness of the problem, the perturbation due to the noise is exceedingly large and the noisy data are not close to the exact data. Similar remarks can be easily extended to the related problems of Abel and Radon transform inversion in Sobolev spaces [34]. In fact they can apply to the general case.

By solving the direct problem for a class X of objects, the linear operator A, defined on X, is determined. Then the set of the functions g in the range of A is the set of the <u>exact data</u> - also called computed data or noise free data. The experimental data, i.e. the results of the experiment, are affected by noise - they are also called noisy data - and can be expressed in the following form

$$g_{\varepsilon} = Af + w_{\varepsilon} \tag{2.7}$$

where ε denotes an estimate of the size of the noise. The first term of the r.h.s. of this equation corresponds to the exact data and the second term is the perturbation due to the noise. Such a perturbation may not be in the range of A: if A is, for instance, an integral operator with an analytic kernel, the exact data are also analytic but the function w_{ε} can be rather irregular. This remark implies that, in general, R(A) is a subset of a space Y which must be sufficiently broad to contain all the experimental data and whose metric must provide a correct description of the effect of noise (the norm of w_{ε} must be of the order of ε). In other words, the inverse problem with noisy data must be formulated as an ill-posed problem.

When $g \notin R(A)$, the equation $Af = g$ has no solution and equation (2.7) is useless because both f and w_{ε} are unknown. It is clear that, in order to master these problems, some drastic modification in the concept of solution is required. To this purpose it is convenient to distinguish two cases:

a) the range of A is closed;

b) the range of A is not closed.

In both cases the solution, when it exists, may not be unique if the null space of A is not trivial.

In the case a), which will be investigated in Sect.3, R(A) is a closed subspace of Y and therefore the solution does not exist when g has a component orthogonal to R(A). Existence for any g is assured if one considers <u>least squares solutions</u>. Furthermore uniqueness can also be assured by looking for a least squares solution of minimal norm, the so-called <u>generalized solution</u>. This procedure is equivalent to projecting the data function onto R(A) and restricting A to the orthogonal complement of N(A). In such a way a generalized inverse of A is defined which is the extension to linear continuous operators of the Moore-Penrose inverse of a matrix [20,39]. When R(A) is closed, the generalized inverse is continuous and therefore the problem of determining the generalized solution is well-posed. However such a problem may be ill-conditioned and sometimes extremely ill-conditioned. Under these circumstances it is necessary to use the same techniques which apply to the case b).

When R(A) is not closed, the generalized inverse of A is not continuous and therefore the problem of determining the generalized solution is still ill-posed. It is easy to see that R(A) is not closed if and only if $\lambda = 0$ is an accumulation point of the spectrum of the operator A*A. Therefore a typical example of an operator in such a class is provided by a compact operator whose rank is not finite.

According to some authors, only problems in this second class deserve the denomination of ill-posed problems. Certainly they are the most difficult ones and precisely for these problems has been developed the so-called <u>regularization theory</u> - see Sects.4,5 - which, roughly speaking, is the theory of the continuous approximations to the discontinuous inverse - or generalized inverse - of the operator A. In its present form this theory is an extension of the method proposed by Tikhonov in the case of first kind Fredholm integral equations [51,52].

3. GENERALIZED INVERSES

The method of generalized solutions provides the tools for overcoming the ill-posedness of problem (1.1) when R(A) is closed. Therefore we will focus on this case but we will also give the main results in the case where R(A) is not closed. Furthermore we will consider an extension of the method which is required in several applications and which can lead to an ill-posed problem even when R(A) is closed.

Examples of operators, whose range is closed and which are of interest in applied inverse problems, are the following:

1) projection operators;
2) finite rank integral operators;
3) convolution operators, $Af = K*f$, when the Fourier (or Mellin) transform of the kernel, $\hat{K}(\xi)$, satisfies the following conditions: i) the support of $\hat{K}$ is bounded; ii) there exist positive constants m, M such that: $m \leq |\hat{K}(\xi)| \leq M$, a.e. on the support of $\hat{K}$;

4) operators of the type $A = \lambda I - K$, where K is a compact operator and λ is an eigenvalue of K (second kind Fredholm integral equations);

5) operators related to inverse problems with discrete data, as defined in Sect.7.

When R(A) is a closed linear subspace of Y, the solution of equation (1.1) does not exist if g is a noisy data with a component orthogonal to R(A). With the help of an example we will show that this is possible in a practical problem as an effect of the noise contribution w_ε, equation (2.7).

Consider the following convolution operator, defined on $X = L^2(\mathbb{R})$ and satisfying the conditions stated at the point 3) above:

$$(Af)(x) = \int_{-\infty}^{+\infty} \frac{\sin[c(x-y)]}{\pi(x-y)} f(y)dy \quad . \tag{3.1}$$

In electrical engineering the system corresponding to such an operator is called a perfect low pass filter and the functions in the range of A are called bandlimited functions. The Fourier transform of these functions, indeed, is zero out of the interval $[-c,c]$ and they form a closed subspace which is a Paley-Wiener space of entire functions of exponential type. Notice that A is just the orthogonal projection of $L^2(\mathbb{R})$ on such a subspace.

Assume now that the experimental data are given by equation (2.7). A typical feature of noise is that the function w_ε may have a Fourier transform which is not zero out of the interval $[-c,c]$. It follows that equation (1.1) has no solution if g is such experimental data. We also notice that uniqueness does not hold true; the null space of A is the subspace of the square integrable functions whose Fourier transform is zero on $[-c,c]$.

3.1 Least squares solutions

If equation (1.1) has no solution, then it is quite natural to look for the function, or the functions, f such that Af is as close as possible to g. Therefore we give the following definition: a function $u \in X$ is said to be a least squares solution (or pseudosolution) of equation (1.1) if it minimizes the distance between Af and g

$$\|Au - g\|_Y = \inf\{\|Af - g\|_Y \mid f \in X\} \quad . \qquad (3.2)$$

It is quite easy to show that the set S_g of least squares solutions, associated with the data g, is just the set of solutions of the Euler equation

$$A^*A\,u = A^*g \qquad (3.3)$$

or also the set of solutions of the equation

$$Au = Pg \qquad (3.4)$$

where P is the projection operator onto R(A). It follows that the determination of least squares solutions is equivalent to solving equation (1.1) with g replaced by the projected data. It is also obvious that, since R(A) is closed, least squares solutions exist for any g and that, when $g \in R(A)$, the set of least squares solutions coincides with the set of solutions of equation (1.1).

If u_0 is a given least squares solution (or solution) associated with the data g, then the set S_g is the closed affine subspace

$$S_g = \{u \in X \mid u = u_0 + v,\ Av = 0\} \qquad (3.5)$$

i.e. S_g is just a translation of N(A), $S_g = u_0 + N(A)$.

3.2 Generalized solutions

For any g, S_g is a closed and convex set and therefore there exists a unique least squares solution of minimal norm. This is called the generalized solution (or also the normal pseudosolution) and is denoted by f^+:

$$\|f^+\|_X = \inf\{\|u\|_X \mid u \in S_g\} \quad . \qquad (3.6)$$

The generalized solution is the unique least squares solution which is orthogonal to N(A) and therefore this procedure is equivalent to restricting the operator A to the orthogonal complement of N(A).

The mapping $g \longrightarrow f^+$ defines a linear operator $A^+: Y \longrightarrow X$ which is called the generalized inverse of A. It is easy to prove that, when A is linear and continuous, A^+ is closed [20]. Now, since R(A) is closed, f^+ exists and is unique for any $g \in Y$ and therefore $D(A^+) = Y$. It follows that A^+ is continuous or, in other words, that the problem of the determination of the generalized solution is well-posed. Its stability is controlled by the condition number

$$\mathrm{cond}(A) = \|A\| \, \|A^+\| \quad . \tag{3.7}$$

If λ_{min} and λ_{max} denote respectively the lower and upper bound of the positive part of the spectrum of the operator AA*, then

$$\mathrm{cond}(A) = (\lambda_{max}/\lambda_{min})^{1/2} \quad . \tag{3.8}$$

One can easily derive this equation from the relationship

$$A^+ = (A^*A)^+A^* = A^*(AA^*)^+ \tag{3.9}$$

and from the spectral representation of AA*.

3.3 Generalized solutions in the case of nonclosed range

When R(A) is not closed, the generalized solution f^+ does not exist for any g. Coming back to Sect.3.1, we see that least squares solutions exist if and only if $Pg \in R(A)$, where P is now the projection operator onto the closure of R(A). If this condition is satisfied, then it is still true that there exists a unique least squares solution of minimal norm. Since the generalized inverse A^+ is a closed operator with domain $D(A^+) = R(A) \oplus R(A)^\perp$, A^+ is not continuous and the problem of determining f^+ is ill-posed.

An important example of operators whose range is not closed is provided by compact operators, on condition that the rank is not finite. If A is such an operator, then the operators A*A and AA* have the same positive eigenvalues with the same multiplicity. Let λ_k be the repeated eigenvalues, ordered in a decreasing sequence ($\lambda_0 \geq \lambda_1 \geq \lambda_2 \geq \ldots$) and let u_k, v_k be the eigenvectors of A*A and AA* respectively, associated with λ_k. It is always possible to choose u_k, v_k in such a way that they solve the coupled equations

$$Au_k = \sqrt{\lambda_k} v_k \quad , \quad A^* v_k = \sqrt{\lambda_k} u_k \quad . \tag{3.10}$$

The positive numbers $\sqrt{\lambda_k}$ are called the singular values and u_k, v_k are called the singular vectors (functions) of the compact operator A. The set of the triples $\{\sqrt{\lambda_k};\ u_k,\ v_k\}$ is the singular system of A. The canonical representation of A in terms of its singular system is

$$Af = \sum_{k=0}^{+\infty} \sqrt{\lambda_k} (f,\ u_k)_X\ v_k \quad . \tag{3.11}$$

It follows that necessary and sufficient conditions for the existence of a solution of equation (1.1) are

$$g \in N(A^*)^{\perp}, \quad \sum_{k=0}^{+\infty} \lambda_k^{-1} \ |(g, v_k)_Y|^2 < +\infty \tag{3.12}$$

which are also known as Picard conditions [44,39]. In particular, the second of the conditions (3.12) is the necessary and sufficient condition for the existence of least squares solutions. Finally, the canonical representation of the generalized inverse is

$$A^+ g = \sum_{k=0}^{+\infty} \lambda_k^{-1/2} \ (g, v_k)_Y \ u_k \tag{3.13}$$

as one can easily derive from (3.11) and the remark that $\{v_k\}$ is a basis in the closure of R(A).

3.4. C-generalized solutions

In some problems where uniqueness does not hold true, the generalized inverse introduced in the previous sections, which is just the extension to linear continuous operators of the Moore-Penrose inverse of a matrix, may not be satisfactory in the sense that the corresponding generalized solutions have unphysical properties. An example will be given in Sect.3.5. For this reason an extension of the generalized solution has been introduced by looking for a least squares solution which minimizes a norm or a seminorm of the following form

$$p(f) = \|Cf\|_Z \tag{3.14}$$

where $C: X \longrightarrow Z$ is a given linear operator from the object space X into a constraint space Z. We assume that Z is also a Hilbert space.

A well known example can be found in interpolation theory: interpolation in terms of spline functions of degree $m = 2k-1$ is obtained by minimizing the seminorm

$$p(f) = \left(\int_a^b |f^{(k)}(x)|^2 \, dx \right)^{1/2} \tag{3.15}$$

where $f^{(k)} = d^k f/dx^k$ [19]. Such a seminorm with $k = 2$ was also used by Phillips [43] for the regularization of first kind Fredholm integral equations, while in the same problem Tikhonov [51,52] has used the norm

$$p(f) = \left(\sum_{k=0}^{m} \int_a^b q_k(x) \, |f^{(k)}(x)|^2 \, dx \right)^{1/2} \tag{3.16}$$

where the $q_k(x)$ are continuous and positive functions. Therefore the previous extension of the generalized solution is also required for investigating the limiting case of certain regularization algorithms (Sect.5.4).

We define the C-generalized solution f_C^+ of equation (1.1) as the least squares solution which minimizes the functional (3.14), i.e.

$$\|Cf_C^+\|_Z = \inf \left\{ \|Cu\|_Z \,\middle|\, u \in S_g \right\} \quad . \tag{3.17}$$

In the case where C is a continuous operator and both R(A) and R(C) are closed, necessary and sufficient conditions on the pair $\{A,C\}$, which assure existence and uniqueness of the C-generalized solution for any $g \in Y$, are given in [39] (pp.234-236). These results, however, do not cover the interesting case of a differential operator. Results which can be applied to this case can be found in [38]. Here we give conditions which are satisfied by the norms and seminorms used in the applications and which assure uniqueness and existence of the C-generalized solution for a suitable set of data functions. These conditions apply to the general case where the null spaces N(A) and N(C) are not trivial.

The basic assumptions on the contraint operator C are the following:

i) $N(A) \cap N(C) = \{0\}$ (uniqueness condition)
ii) $C: X \longrightarrow Z$ is a closed operator with D(C) dense in X and R(C) = Z;
iii) A N(C) is closed in Y.

Notice that condition iii) is trivially satisfied in the case of seminorms defined in terms of differential operators since in that case N(C) is a finite dimensional subspace of X (and A is continuous).

Conditions i)-iii) imply that there exists a constant $\beta > 0$ such that for any $f \in D(C)$

$$\beta \|f\|_X^2 \leqslant \|Af\|_Y^2 + \|Cf\|_Z^2 \quad . \tag{3.18}$$

This condition is called by Morozov a "completion condition" [38].

In order to prove the inequality (3.18), we introduce the space $W = Y \oplus Z$ and the operator $B: X \longrightarrow W$ defined as follows

$$Bf = \{Af,\ Cf\} \quad , \quad f \in D(C) \quad . \tag{3.19}$$

Then, since B is closed, inequality (3.18) is equivalent to the statement that B has an inverse B^{-1} and that R(B) is closed in W.

The existence of B^{-1} is a trivial consequence of condition i). In order to prove that R(B) is closed, let $\{f_n\} \subset D(C)$ be a sequence such that $\{Bf_n\}$ is convergent in W. Then $\{Af_n\}$ is convergent in Y and also $\{Cf_n\}$ is convergent in Z. Put $f_n = f_n^{(0)} + f_n^{(1)}$ with $f_n^{(0)} \in N(C)$ and $f_n^{(1)} \in N(C)^{\perp}$. Since R(C) = Z, the restriction of C to $N(C)^{\perp}$ has a bounded inverse and therefore there exists a constant $\gamma > 0$ such that

$$\|Cf_n\|_Z = \|Cf_n^{(1)}\|_Z \geqslant \gamma \|f_n^{(1)}\|_Z \quad . \tag{3.20}$$

This implies that $\{f_n^{(1)}\}$ is also convergent and, thanks to the closure of C, its limit $f^{(1)}$ belongs to D(C) and $Cf_n \longrightarrow Cf^{(1)}$. Now we have

$Af_n = Af_n^{(0)} + Af_n^{(1)}$ and both $\{Af_n\}$ and $\{Af_n^{(1)}\}$ are convergent. It follows that $\{Af_n^{(0)}\}$ is also convergent and, thanks to the closure of $AN(C)$, there exists $f^{(0)} \in N(C)$ such that $Af_n^{(0)} \longrightarrow Af^{(0)}$. By combining all the results we have $Bf_n \longrightarrow B(f^{(0)}+f^{(1)})$ and therefore $R(B)$ is closed.

Condition (3.18) implies the following result [38]:

<u>Theorem 3.1</u> - For any g such that $Pg \in AD(C)$, where P is the projection operator onto the closure of $R(A)$, there exists a unique C-generalized solution f_C^+.

<u>Proof</u>: Let S_g be the set of least squares solutions associated with g. If $Pg \in AD(C)$, then $S_g \cap D(C)$ is not empty and convex. Furthermore we put

$$\mu_g = \inf\{\|Af - g\|_Y \mid f \in D(C)\} \tag{3.21}$$

$$\nu_g = \inf\{\|Cu\|_Z \mid u \in S_g \cap D(C)\} \quad . \tag{3.22}$$

First we prove <u>uniqueness</u>. Let $u_1, u_2 \in S_g \cap D(C)$ be two least squares solutions such that $\|Cu_1\|_Z = \|Cu_2\|_Z = \nu_g$. From the parallelogram law we have

$$\begin{aligned} \left\|\tfrac{1}{2} A(u_1-u_2)\right\|_Y^2 &= \tfrac{1}{2} \|Au_1 - g\|_Y^2 + \tfrac{1}{2} \|Au_2 - g\|_Y^2 - \\ &- \left\|\tfrac{1}{2} A(u_1+u_2) - g\right\|_Y^2 = \tfrac{1}{2} (\mu_g^2+\mu_g^2) - \mu_g^2 = 0 \end{aligned} \tag{3.23}$$

$$\begin{aligned} \left\|\tfrac{1}{2} C(u_1-u_2)\right\|_Z^2 &= \tfrac{1}{2} \|C u_1\|_Z^2 + \tfrac{1}{2} \|Cu_2\|_Z^2 - \\ &- \left\|\tfrac{1}{2} C(u_1+u_2)\right\|_Z = \nu_g^2 - \left\|\tfrac{1}{2} C(u_1+u_2)\right\|_Z^2 \leq 0 \end{aligned} \tag{3.24}$$

and therefore $A(u_1-u_2) = 0$, $C(u_1-u_2) = 0$. Then condition i) implies $u_1 = u_2$.

As concerns <u>existence</u>, let $\{u_n\} \subset S_g \cap D(C)$ be a minimizing sequence, so that

$$Au_n = Pg \quad , \quad \lim \|Cu_n\|_Z = \nu_g \quad . \tag{3.25}$$

From the parallelogram law it follows that $\{Cu_n\}$ is a Cauchy sequence in Z and this result, combined with the first of eqs.(3.25) and the inequality (3.18), implies that $\{u_n\}$ is also a Cauchy sequence in X. Let $u^{(0)}$ be the limit. Since C in closed, $u^{(0)} \in D(C)$ and $Cu_n \longrightarrow Cu^{(0)}$ so that, using also the continuity of A and eqs.(3.25)

$$Au^{(0)} = Pg \quad , \quad \| Cu^{(0)} \|_Z = \nu_g \quad . \tag{3.26}$$

These conditions just imply that $u^{(0)} = f_C^+$. □

The mapping $g \longrightarrow f_C^+$ defines a linear operator $A_C^+ : Y \longrightarrow X$ which will be called the <u>C-generalized inverse</u> of A. In the case where C has a bounded inverse C^{-1}, it is possible to find a very simple representation of A_C^+. If we put $f = C^{-1}\phi$, we find that the determination of the C-generalized inverse of A is equivalent to the determination of the generalized inverse of AC^{-1} and the final result is

$$A_C^+ = C^{-1}(AC^{-1})^+ \quad . \tag{3.27}$$

Finally we point out that, even when R(A) is closed, the <u>problem of the determination of the C-generalized solution may be ill-posed</u>. This clearly appears in the case (3.27). If C^{-1} is compact, as it may be when C is a differential operator, then AC^{-1} is also compact and therefore its generalized inverse is not continuous. Then it is necessary to use regularization techniques, which will be described in the next sections, in order to find continuous approximations of the C-generalized solution.

3.5 An example: the perfect low pass filter

A simple example of an application of the previous theory is provided by the inversion of the operator (3.1), the perfect low pass filter, which is a projection operator in $X = Y = L^2(\mathbb{R})$.

For a given $g \in L^2(R)$ the set of least squares solutions can be easily characterized. Let $\hat{g}$ be the Fourier transform of g and let χ_c be the characteristic function of the interval $[-c,c]$. Then the set of the least squares solutions is the set of all the functions given by

$$\hat{u}(\xi) = \chi_c(\xi)\,\hat{g}(\xi) + [1-\chi_c(\xi)]\,\hat{\psi}(\xi) \tag{3.28}$$

where $\hat{\psi}$ is an arbitrary square integrable function. Then the Fourier transform of the generalized solution is

$$\hat{f}^+(\xi) = \chi_c(\xi)\,\hat{g}(\xi) \tag{3.29}$$

and therefore

$$f^+(x) = \frac{1}{2\pi}\int_{-\infty}^{+\infty} \frac{\sin[c(x-y)]}{\pi(x-y)}\, g(y)\, dy \quad . \tag{3.30}$$

Notice that f^+ is just the projection of the data function g onto R(A) and this property holds true whenever A is a projection operator.

The generalized solution (3.30) may be unsatisfactory if, for physical reasons, the Fourier transform of the solution must be continuous or differentiable. In such a case it is quite natural to introduce a C-generalized solution by looking for the least squares solution which minimizes the norm ($Z = X = L^2(\mathbb{R})$)

$$\|Cf\|_Z^2 = \int_{-\infty}^{+\infty} (1+x^2)\,|f(x)|^2\, dx = \tag{3.31}$$

$$= \frac{1}{2\pi}\int_{-\infty}^{+\infty} |\hat{f}(\xi)|^2\, d\xi + \frac{1}{2\pi}\int_{-\infty}^{+\infty} |\hat{f}'(\xi)|^2\, d\xi \quad .$$

The set $S_g \cap D(C)$ is not empty if and only if the restriction of $\hat{g}$ to $[-c,c]$ has a square integrable derivative. When this condition is satisfied the set $S_g \cap D(C)$ is the set of all the functions (3.28) with $\hat{\psi}$ satisfying the conditions

$$\int_{|\xi|>c} |\hat{\psi}'(\xi)|^2\, d\xi < +\infty \quad , \quad \hat{\psi}(\pm c) = \hat{g}(\pm c) \quad . \tag{3.32}$$

Then the function $\hat{\psi}$ which minimizes (3.31) is

$$\hat{\psi}(\pm\xi) = \hat{g}(\pm c)\, \exp[-(\xi-\pi)] \; , \; \xi > \pi \quad . \tag{3.33}$$

It is obvious that the problem of determining such a C-generalized solution is ill-posed in $L^2(\mathbb{R})$.

In several practical problems it is also known that the solution has a bounded support, say $[-1,1]$, so that its Fourier transform is analytic.

In such a case we can consider the restriction of the operator (3.1) to $L^2(-1,1)$, i.e.

$$(Af)(x) = \int_{-1}^{1} \frac{\sin[c(x-y)]}{\pi(x-y)} f(y)dy \quad . \tag{3.34}$$

This is a compact operator from $L^2(-1,1)$ into $L^2(\mathbb{R})$ and it has an inverse. The equation $Af = 0$, indeed, has only the solution $f = 0$: the Fourier transform $\hat{f}$ of any solution of this equation is zero over $(-c,c)$ and, since it is analytic, it is zero everywhere.

Therefore uniqueness holds true but the problem is still ill-posed because A^{-1} is not continuous. The singular functions of A are the <u>linear prolate spheroidal functions</u> investigated by Slepian and coworkers [49].

4. CONSTRAINED LEAST SQUARES SOLUTIONS

If R(A) is not closed, the generalized inverse is not continuous. When $Pg \notin R(A)$ the equation $Af = Pg$ has no solution, but it is always possible to find a sequence $\{f_n\} \subset N(A)$ such that $Af_n \longrightarrow Pg$, $n \longrightarrow \infty$. In this case, $\|f_n\|_X \longrightarrow \infty$. On the other hand, when $Pg \in R(A)$, the generalized solution f^+ exists but a small variation of g can produce an arbitrarily large variation of f^+, since one can always find a sequence $\{f_n\} \subset N(A)$ such that $\|Af_n\|_Y \longrightarrow 0$ and $\|f_n\|_X \longrightarrow \infty$. A classical example of such a sequence is that given by Hadamard [22] in the case of the Cauchy problem for the Laplace equation.

In other words, for a given $\varepsilon > 0$ and a given $g \in Y$, the set

$$J_\varepsilon(g) = \{ f \in N(A)^\perp \mid \|Af - g\|_Y \leqslant \varepsilon \} \tag{4.1}$$

is not bounded in X.

Under these circumstances it seems meaningless to look for approximate solutions, i.e. for functions f such that Af is close to Pg. Nevertheless, in a practical problem, methods for producing reasonable approximate solutions are required.

The basic idea in some pioneering works on ill-posed problems for partial differential equations [45,28,15,29] was to look for approximate solutions satisfying prescribed bounds. One of the main results was the proof of their continuous dependence on the data in the sense that, when the error on the data tends to zero, these approximate solutions converge to the true solution of the problem. Furthermore it was recognized that a sufficient condition for assuring continuity is the compactness of the set where one is looking for the approximate solutions. An important preliminary result along these lines is a topological lemma, due to Tikhonov [53,33] which in our context can be formulated as follows:

Lemma 4.1 - Let H be a compact subset of the Hilbert space X and assume that the linear, continuous operator $A: X \longrightarrow Y$, when restricted to H, has an inverse. Then the inverse is continuous.

The proof of the Lemma just follows from the remark that the linear continuous operator A maps closed subsets of H into closed subsets of AH. Furthermore the Lemma can be extended to the case of the weak topology. It follows that, if the set H is closed and bounded and the other conditions of the Lemma are satisfied, then the inverse operator is weakly continuous.

4.1 Least squares solutions in compact sets

Assume that, on the basis of physical properties of the solution, it is possible to identify a compact subset of X which contains the unknown solution of the problem. Then it is quite natural to look for functions $\tilde{f} \in H$ such that $A\tilde{f}$ is as close as possible to the data function g, i.e. to solve the problem

$$\|A\tilde{f} - g\|_Y = \inf\{\|Af - g\|_Y \mid f \in H\} \quad . \tag{4.2}$$

Thanks to the compactness of H, such a problem has always a solution which will be called H-constrained least squares solution. Sufficient conditions for uniqueness are provided in the next Theorem [28].

Theorem 4.1 - If the compact set H is convex and the restriction of A to H admits an inverse operator, then, for any $g \in Y$, there exists a unique H-constrained least squares solution $\tilde{f}$. Furthermore the mapping $g \longrightarrow \tilde{f}$ is continuous.

Proof: Since H is compact and A is continuous, the set $AH \subset Y$ is closed. It is also convex since H is convex and A is linear. Therefore, for any $g \in Y$ there exists a unique convex projection $\tilde{g}$ of g onto AH, which is the solution of the problem

$$\|\tilde{g} - g\|_Y = \inf\{\|\bar{g} - g\|_Y \mid \bar{g} \in AH\} \quad . \tag{4.3}$$

Since $\tilde{g} \in AH$ and the inverse of A, restricted to H, exists, there exists a unique solution $\tilde{f} \in H$ of the equation $A\tilde{f} = \tilde{g}$. From (4.3) it follows that $\tilde{f}$ is the unique solution of (4.2). Finally, the continuity of the mapping $g \longrightarrow \tilde{f}$ is a consequence of the following remarks: i) the mapping $g \longrightarrow \tilde{g}$ is continuous since the following inequality holds true:

$$\|\tilde{g}_2 - \tilde{g}_1\|_Y \leq \|g_2 - g_1\|_Y \tag{4.4}$$

where $\tilde{g}_1$, $\tilde{g}_2$ are the convex projections of g_1, g_2 respectively; ii) the mapping $\tilde{g} \longrightarrow \tilde{f}$ is continuous as a consequence of Lemma 4.1. □

From the previous Theorem it follows that, if we have data functions g_ε converging to $g \in AH$, $\|g_\varepsilon - g\|_Y \leq \varepsilon$, then the corresponding H-constrained least squares solutions $\tilde{f}_\varepsilon$ converge to the unique solution of the problem: $Af = g$, $f \in H$. In other words, if the errors on the data tend to zero and if the compact and convex set has been correctly chosen, then the H-constrained least squares solutions provide an accurate approximation of the solution of the problem.

We also notice that, in the case where H is closed and bounded, the previous theorem still holds true but the mapping $g \longrightarrow \tilde{f}$ in general is only weakly continuous.

According to Theorem 4.1, the uniqueness of the constrained least squares solution derives both from the convexity of the set H and from the existence of the inverse of the operator A. The latter condition can sometimes be removed when H is the level set of a convex functional of the type (3.14), i.e. the set of all the functions $f \in D(C)$

satisfying the conditon $p(f) \leq E$, where E is a prescribed constant. We first consider in detail the case where the constraint operator C is the identity operator in X and therefore the set H is a sphere in X:

$$H = \{ f \in X \mid \|f\|_X \leq E \} \quad . \tag{4.5}$$

The next Lemma contains results which will be used in this and in the next Section.

Lemma 4.2 - For any $g \in Y$ and $\alpha > 0$, there exists a unique $f_\alpha \in X$ which minimizes the functional

$$\Phi_\alpha[f] = \|Af - g\|_Y^2 + \alpha \|f\|_X^2 \quad . \tag{4.6}$$

It is the unique solution of the Euler equation

$$(A^*A + \alpha I) f_\alpha = A^*g \tag{4.7}$$

and furthermore $f_\alpha \in N(A)^\perp$. The function

$$\rho(\alpha) = \|Af_\alpha - g\|_Y \tag{4.8}$$

is an increasing function of α with values in $(\|Qg\|_Y, \|g\|_Y)$, where $Q = I - P$ is the orthogonal projection onto $R(A)^\perp$, while the function

$$\nu(\alpha) = \|f_\alpha\|_X \tag{4.9}$$

is a decreasing function of α with values in $(\|A^+g\|_X, 0)$ - in $(+\infty, 0)$ when $Pg \notin R(A)$.

Proof: Any function which minimizes the functional (4.6) is also a solution of the Euler equation (4.7). The latter has a unique solution for any $g \in Y$ and $\alpha > 0$ since the operator $A^*A + \alpha I$ is positive definite and therefore has a continuous inverse. The property $f_\alpha \in N(A)^\perp$ follows from the relation

$$(A^*A + \alpha I)^{-1} A^* = A^*(AA^* + \alpha I)^{-1} \tag{4.10}$$

since this implies $f_\alpha \in R(A^*)$. Notice that the I of the l.h.s. denotes the identity operator of X, while the I of the r.h.s. denotes the identity operator of Y. Using this relation and the property $A^*g = A^*Pg$, one gets the following expressions for the functions $\rho(\alpha)$, $\nu(\alpha)$:

$$\rho^2(\alpha) = \alpha^2 \|(AA^* + \alpha I)^{-1}Pg\|_Y^2 + \|Qg\|_Y^2 \tag{4.11}$$

$$\nu^2(\alpha) = (AA^*(AA^* + \alpha I)^{-1} Pg, (AA^* + \alpha I)^{-1} Pg)_Y \tag{4.12}$$

and the properties of these functions stated in the Lemma follow from the spectral representation of the self-adjoint operator AA^*. □

The next Theorem is an extension of a result proved in [26] in the case of a compact operator A.

Theorem 4.2 - Let H be the sphere (4.5). Then, if $\|A^+g\|_X \leq E$, the set of the H-constrained least squares solutions is $S_g \cap H$ and therefore $f^+ = A^+g$ is the unique element of minimal norm of this set. On the other hand, if $\|A^+g\|_X > E$ (in particular, $= +\infty$), then there exists a unique solution of the problem

$$\|A\tilde{f} - g\|_Y = \inf\{\|Af - g\|_Y \mid \|f\|_X \leq E\} \tag{4.13}$$

which is given by the unique function f_α, the solution of eq.(4.7), satisfying the condition $\nu(\alpha) = E$.

Proof: The first part of the Theorem is obvious. On the other hand, when the condition $\|A^+g\|_X \leq E$ is not satisfied, either the set of the (unconstrained) least squares solutions does not intersect the sphere H or the set is empty because $Pg \notin R(A)$. In both cases, if the functional $\Psi[f] = \|Af - g\|_Y^2$ has a minimum point in H, this point must lie on the boundary of H, i.e. must satisfy the condition $\|\tilde{f}\|_X = E$. It follows that one can use the method of Lagrange multipliers for determining the minimum of $\Psi[f]$ in H: minimize the functional (4.6) for any $\alpha > 0$ and look for the minimum point f_α which satisfies the condition $\nu(\alpha) = E$. Since, as proved in Lemma 4.2, $\nu(\alpha)$ is a decreasing function with $\nu(0+) = \|A^+g\|_X > E$ and $\nu(+\infty) = 0$, there exists a unique value of α satisfing the prescribed condition. □

Since the sphere is weakly compact in X, at a glance it seems possible to state only the weak continuity of the mapping $g \longrightarrow \tilde{f}$, $\tilde{f}$ being the solution of problem (4.13). However, the fact that, whenever $\|A^+g\|_X > E$, $\tilde{f}$ lies on the boundary of the sphere, has some important implications. Let g_ε be a noisy data function such that $\|g_\varepsilon - g\|_Y \leq \varepsilon$ where $g \in R(A)$ is the exact data function. Assume also that $\|A^+g_\varepsilon\|_X > E$ and that this condition is always satisfied when $\varepsilon \longrightarrow 0$, so that the corresponding constrained least squares solutions $\tilde{f}_\varepsilon$ satisfy the condition $\|\tilde{f}_\varepsilon\|_X = E$. Then we have the following cases:

i) if $\|A^+g\|_X < E$, i.e. the prescribed constant E is overestimated, then $\tilde{f}_\varepsilon$ weakly converges to f^+, the generalized solution associated with g (the result just follows from the remark that $\tilde{f}_\varepsilon \in N(A)^\perp$);

ii) if $\|A^+g\|_X = E$, i.e. the prescribed constant is precise, then $\tilde{f}_\varepsilon$ strongly converges to f^+ (the result follows from i) and the equality $\|\tilde{f}_\varepsilon\|_X = E = \|f^+\|_X$);

iii) if $\|A^+g\|_X > E$, i.e. the prescribed constant is underestimated, then $\tilde{f}_\varepsilon$ strongly converges to the H-constrained least squares solution associated with the exact data g.

In the case where A is a compact operator, the solution of problem (4.13) has the following representation in terms of the singular system of A:

$$\tilde{f} = \sum_{k=0}^{+\infty} \lambda_k^{1/2} (\lambda_k + \alpha)^{-1} (g, v_k)_Y u_k \tag{4.14}$$

where α is the unique solution of the equation

$$\sum_{k=0}^{+\infty} \lambda_k (\lambda_k + \alpha)^{-2} |(g, v_k)_Y|^2 = E^2 \quad . \tag{4.15}$$

In the particular case where A is an integral operator with an analytic kernel, then $\tilde{f}$ is also analytic. It is sufficient to notice that $\tilde{f} \in R(A^*)$ and that A^* is also an integral operator with analytic kernel.

A few remarks on the more general case where H is a level set of a functional of the type (3.14) follow. We consider

$$H = \{ f \in D(C) \mid \|Cf\|_Z \leq E \} \quad . \qquad (4.16)$$

For simplicity we assume that the closed operator C has a bounded inverse C^{-1}. Then the set H is closed and bounded, since $H = C^{-1} \mathcal{S}_E$ where $\mathcal{S}_E$ is the closed sphere of radius E in Z. In general H is only weakly compact; it is compact if and only if C^{-1} is compact.

Then the problem of determining $\tilde{f}$ such that

$$\|A\tilde{f} - g\|_Y = \inf \{ \|Af - g\|_Y \mid \|Cf\|_Z \leq E \} \qquad (4.17)$$

can be reduced to the problem (4.13) just by putting $\phi = Cf$, $\bar{A} = A C^{-1}$. It follows that $\tilde{f} = C^{-1}\tilde{\phi}$, with $\tilde{\phi}$ given by Theorem 4.2, and $C\tilde{f} = \tilde{\phi} \in R(\bar{A}^*) \subset D(C^*)$ (notice that, since D(C) and D(C*) are dense in X and Z respectively, then $(C^{-1})^* = (C^*)^{-1}$). This implies that $\tilde{f} \in D(C^*C)$. Therefore $\tilde{f}$ can be obtained by solving for any α the equation

$$(A^*A + \alpha C^*C) f_\alpha = A^*g \qquad (4.18)$$

and then looking for the unique f_α satisfying the condition

$$\|Cf_\alpha\|_Z = E \quad . \qquad (4.19)$$

As concerns the continuity of the mapping $g \longrightarrow \tilde{f}$, the conclusions derived in the case C = I apply also to this case, since they apply to the mapping $g \longrightarrow \tilde{\phi}$ and $\tilde{f}$ is obtained from $\tilde{\phi}$ by means of the continuous operator C^{-1}. It is obvious that, when C^{-1} is compact, strong continuity always holds true.

4.2 Least squares methods using a bound on the error

The previous approach requires some a priori information about the solution f, namely that it belongs to a prescribed compact (or weakly compact) set of the object space X . A method for the solution of ill-posed problems proposed in [36] requires a prescribed bound both on the solution and on the error. More precisely it is assumed that positive constants ε, E are known such that $f \in K = H \cap J_\varepsilon(g)$, where H is given by eq.(4.16) and $J_\varepsilon(g)$ by eq.(4.1). By means of a scaling ($f/E \longrightarrow f$, $g/E \longrightarrow g$, $\varepsilon/E \longrightarrow \varepsilon$) it is always possible to have $E = 1$, so that the set K is given by

$$K = \left\{ f \in D(C) \mid \|Af - g\|_Y \leq \varepsilon \, , \ \|Cf\|_Z \leq 1 \right\} \quad . \tag{4.20}$$

If K is not empty, then any $f \in K$ is an admissible approximate solution of the problem.

It is easily recognized that K is not empty if and only if we have $\|A\tilde{f}-g\|_Y \leq \varepsilon$, where $\tilde{f}$ is the constrained least squares solution satisfying the condition $\|C\tilde{f}\|_Z = 1$. A sufficient condition can be obtained as follows. We consider the functional

$$\Phi_0[f] = \|Af - g\|_Y^2 + \varepsilon^2 \|Cf\|_Z^2 \tag{4.21}$$

and we consider the two sets [50,6]

$$K_0 = \left\{ f \in D(C) \mid \Phi_0[f] \leq \varepsilon^2 \right\} \tag{4.22}$$

$$K_1 = \left\{ f \in D(C) \mid \Phi_0[f] \leq 2\varepsilon^2 \right\} \quad . \tag{4.23}$$

Let $\tilde{f}_0$ be the minimum point of Φ_0; under the assumptions made in the previous section, $\tilde{f}_0$ is the unique solution of the Euler equation (4.19) with $\alpha = \varepsilon^2$. Then, if the following compatibility condition

$$\Phi_0[\tilde{f}_0] \leq \varepsilon^2 \tag{4.24}$$

is satisfied, the set K_0 is not empty and also the sets K, K_1 are not empty, since the following inclusions hold true

$$K_0 \subset K \subset K_1 \tag{4.25}$$

(the second inclusion will be used in Sect.6 for obtaining an upper bound on the diameter of K). Notice that it has also been obtained explicitly an element of the set K, i.e. an approximate solution compatible with the prescribed constraints. The advantage with respect to the method of Sect. 4.1 is obvious: in that case one has to solve equation (4.18) for several values of α and then determine the value of α which satisfies condition (4.19). In the present case one has to solve equation (4.18) just for one value of α , namely $\alpha = \varepsilon^2$, and then verify that the compatibility condition (4.24) is satisfied.

As concerns the convergence of these least squares solutions when the error on the data tends to zero, we first consider the case C = I. Then results analogous to those discussed in Sect. 4.1 can be proved [6], provided that the compatibility condition (4.24) is satisfied.

Let g_ε be data functions affected by errors and such that $\|g_\varepsilon - g\|_Y \leqslant \varepsilon$, where $g \in R(A)$. Furthermore, let f^+ be the generalized solution associated with g and let $\tilde{f}_{0,\varepsilon}$ be the function which minimizes the functional (4.21) with C = I and g replaced by g_ε . The following result holds true.

Theorem 4.3 - If, for any ε the compatibility condition (4.24) is satisfied, then, when $\varepsilon \longrightarrow 0$, $\tilde{f}_{0,\varepsilon}$ weakly converges to f^+. The convergence is strong if the prescribed constant is precise, i.e. $\|f^+\|_X = 1$.

Proof: From the compatibility condition it follows that, for any ε, $\|\tilde{f}_{0,\varepsilon}\|_X \leqslant 1$. Therefore it is possible to extract a weakly convergent subsequence. Let f_0 be the limit. Since $\tilde{f}_{0,\varepsilon} \in N(A)^\perp$ (see Lemma 4.2), then $f_0 \in N(A)^\perp$. Furthermore, the compatibility condition implies also that $\|A\tilde{f}_{0,\varepsilon} - g_\varepsilon\|_Y \leqslant \varepsilon$ and, by the triangle inequality, $\|A\tilde{f}_{0,\varepsilon} - g\|_Y \leqslant 2\varepsilon$. It follows $A\tilde{f}_{0,\varepsilon} \longrightarrow g$ and, from the previous result, $A\tilde{f}_{0,\varepsilon} \longrightarrow Af_0$. Therefore: $Af_0 = g$, $f_0 \in N(A)^\perp$, i.e. $f_0 = f^+$. Since any subsequence converges to f^+, we have that $\tilde{f}_{0,\varepsilon}$ weakly converges to f^+.

On the other hand, if the prescribed constant is precise, from the weak lower semicontinuity of the norm it follows

$$1 = \|f^+\|_X \leqslant \lim.\inf. \|\tilde{f}_{0,\varepsilon}\|_X \qquad (4.26)$$

and this inequality, combined with

$$\lim.\sup. \|\tilde{f}_{0,\varepsilon}\|_X \leqslant 1 \qquad (4.27)$$

implies $\lim \|\tilde{f}_{0,\varepsilon}\|_X = 1 = \|f^+\|_X$ and also the strong convergence. □

The case of a more general constraint operator can be reduced to the previous one, as in Sect.4.1, provided that C has a bounded inverse C^{-1}. Notice again that the convergence is always strong if C^{-1} is compact.

As a final remark we point out that, as has been shown in [5,50], the method described above is formally quite similar to a probabilistic approach to the solution of ill-posed problems, known as the Wiener filter method [16].

5. REGULARIZATION ALGORITHMS

Some of the methods investigated in the previous Section can be embedded in a more general approach called by Tikhonov the regularization method [51,52] . It consists in the introduction of families of continuous approximations to the discontinuous inverse, or generalized inverse, of the operator A.

As has been shown in Sect.3, when the null space of A is not trivial one can introduce the generalized inverse or also various C-generalized inverses of A. The latter may be discontinuous even when R(A) is closed. Therefore for each of them one can introduce regularization methods. However we will first focus on the (Moore-Penrose) generalized inverse and only in a second stage we will consider C-generalized inverses.It is important to point out that all the methods considered in this section apply also to the case where the operator A has an inverse, since in that case the C-generalized inverses are simply restrictions of A^{-1}.

A one parameter family of operators $\{R_\alpha\}_{\alpha>0}$ is called a regularization algorithm or a regularizer for the solution or generalized solution of equation (1.1), if the following conditions are satisfied [33,53]:

i) for any $\alpha>0$, $R_\alpha : Y \longrightarrow X$ is continuous;
ii) for any g such that $Pg \in R(A)$

$$\lim_{\alpha \downarrow 0} \|R_\alpha g - f^+\|_X = 0 \qquad (5.1)$$

where f^+ is the generalized solution (or solution when A^{-1} exists) of equation (1.1).

When R_α is linear we have a linear regularization algorithm; the variable α is called the regularization parameter.

Equation (5.1) can also be written as follows

$$\lim_{\alpha \downarrow 0} \|R_\alpha A f^+ - f^+\|_X = 0 \quad . \tag{5.2}$$

Therefore the operator $T_\alpha : X \longrightarrow X$, defined by

$$T_\alpha = R_\alpha A \tag{5.3}$$

is an approximation of the orthogonal projection onto $N(A)^\perp$ or of the identity operator when $N(A) = \{0\}$. In the cases where T_α is an integral operator, its kernel can be called, using the language of electrical engineering, the impulse response of the system consisting of the instrument for the determination of the data fucntion g (in the absence of noise) plus the computer using the algorithm R_α.

Conditions i), ii) imply that, for any α and for any exact data $g \in R(A)$, $R_\alpha g$ is a continuous approximation of the solution or generalized solution of eq.(1.1). However, the important case is that of noisy data g_ε, with $Pg_\varepsilon \notin R(A)$, which are close to exact data $g \in R(A)$, $\|g_\varepsilon - g\|_Y \leq \varepsilon$. In such a case no solution of the equation $Af = g_\varepsilon$ exists. Then, if we consider the functions $f_\alpha = R_\alpha g_\varepsilon$, it is easy to see that there must exist an optimum value of α such that f_α is as close a possible to f^+, the generalized solution associated with the exact data g.

Assuming that R_α is linear and that the noisy data g_ε are written in the form (2.8), we get

$$R_\alpha g_\varepsilon - f^+ = (R_\alpha A f^+ - f^+) + R_\alpha w_\varepsilon \tag{5.4}$$

and therefore

$$\|R_\alpha g_\varepsilon - f^+\|_X \leq \omega(\alpha;\, f^+) + \varepsilon N(\alpha) \tag{5.5}$$

where

$$N(\alpha) = \| R_\alpha \| \quad , \quad \omega(\alpha; f^+) = \| R_\alpha A f^+ - f^+ \|_X \quad . \tag{5.6}$$

The first term of the r.h.s. of eq.(5.5) represents the approximation error introduced by the choice of a non-zero value of the regularization parameter; it tends to zero when $\alpha \longrightarrow 0$. The second term is an estimate of the error on the approximate solution induced by the error on the data; it tends to infinity when $\alpha \longrightarrow 0$. Therefore it is necessary to find a compromise between approximation and error magnification. Assume, for simplicity that $N(\alpha)$ and $\omega(\alpha;f^+)$ are monotone functions of α - this condition is satisfied by all the regularizing algorithms used in practice - and, more precisely, that $N(\alpha)$ is a decreasing function, with $N(0+) = +\infty$, while $\omega(\alpha;f^+)$ is an increasing function, with $\omega(0+; f^+) = 0$. Under these assumptions, there exists a unique value of α, $\alpha(\varepsilon)$, which minimizes the r.h.s. of eq.(5.5) and which represents the optimum compromise between approximation and error magnification. Furthermore $\alpha = \alpha(\varepsilon) \longrightarrow 0$, when $\varepsilon \longrightarrow 0$, and $R_\alpha g_\varepsilon \longrightarrow f^+$.

The previous argument implies that a regularization algorithm can give approximate and stable solutions which converge to the exact solution when the error on the data tends to zero. The practical inconvenience is that this method for determining the regularization parameter requires the knowledge of f^+. This difficulty can be partially removed if it is possible to estimate $\omega(\alpha;f^+)$ for a suitable class of solutions, but we will discuss this point in Sect.6.

5.1 The Tikhonov regularizer and the discrepancy principle

The most intensively investigated example of a regularization algorithm is the so-called Tikhonov regularizer, given by

$$R_\alpha = (A^*A + \alpha I)^{-1} A^* \quad . \tag{5.7}$$

Its remarkable properties derive from the fact that it can be obtained by minimizing the functional (4.6) - see Lemma 4.1. In particular, from eq.(4.10) it follows that $R_\alpha g \in N(A)^\perp$, for any $g \in Y$ and this implies that $\{R_\alpha\}_{\alpha>0}$ is a regularization algorithm for A^+. Using the spectral representation of A*A, indeed, it is easy to show that

$$\omega(\alpha;f^+) = \|R_\alpha Af^+ - f^+\|_X = \alpha\|(A^*A + \alpha I)^{-1} f^+\|_X \tag{5.8}$$

tends to zero when $\alpha \longrightarrow 0$. Furthermore, for any f^+, $\omega(\alpha;f^+)$ is an increasing function of α.

It is also easy to find an estimate for $N(\alpha)$, eq.(5.6). Using equation (4.10) we have

$$\|R_\alpha g\|_X^2 = (AR_\alpha g,\ (A^*A + \alpha I)^{-1}\ g)_Y \leqslant \frac{1}{\alpha}\ \|g\|^2{}_Y \tag{5.9}$$

since $\|AR_\alpha\| \leqslant 1$, and therefore

$$N(\alpha) \leqslant \frac{1}{\sqrt{\alpha}} \quad . \tag{5.10}$$

From this bound and equation (5.5) one can derive results concerning possible choices of the function $\alpha(\varepsilon)$.

<u>Theorem 5.1</u> - Let $\alpha = \alpha(\varepsilon)$ be any function such that

$$\varepsilon^2/\beta_1(\varepsilon) \leqslant \alpha(\varepsilon) \leqslant \beta_2(\varepsilon) \tag{5.11}$$

where β_1, β_2 are arbitrary continuous functions, tending to zero when $\varepsilon \longrightarrow 0$. Then $R_\alpha g_\varepsilon$ strongly converges to f^+, the generalized solution associated with the exact data g. On the other hand, if $\alpha = \alpha(\varepsilon)$ is any function such that

$$c_1\ \varepsilon^2 \leqslant \alpha(\varepsilon) \leqslant c_2\ \varepsilon^2 \tag{5.12}$$

where c_1, c_2 are arbitrary constants, then $R_\alpha g_\varepsilon$ weakly converges to f^+.

<u>Proof</u>: As concerns the first part of the Theorem, from inequalities (5.5), (5.10), (5.11) we have

$$\|R_\alpha g_\varepsilon - f^+\|_X \leqslant \omega(\alpha\ ;\ f^+) + \frac{\varepsilon}{\sqrt{\alpha}} \leqslant \omega(\alpha\ ;\ f^+) + \sqrt{\beta_1(\varepsilon)} \tag{5.13}$$

and therefore the r.h.s. tends to zero since $\beta_1(\varepsilon) \longrightarrow 0$ and $\alpha = \alpha(\varepsilon) \longrightarrow 0$.

As concerns the second part of the Theorem, since $f_\alpha = R_\alpha g_\varepsilon$ minimizes the functional (4.6), we have

$$\|Af_\alpha - g_\varepsilon\|_Y^2 + \alpha\|f_\alpha\|_X^2 \leqslant \tag{5.14}$$

$$\leqslant \|Af^+ - g_\varepsilon\|_Y^2 + \alpha\|f^+\|_X^2 \leqslant \varepsilon^2(1+c_2\|f^+\|_X^2) \quad .$$

Also, from the lower bound (5.12) it follows that

$$\|Af_\alpha - g_\varepsilon\|^2{}_Y + \alpha\|f_\alpha\|^2{}_X \geqslant \tag{5.15}$$

$$\geqslant \|Af_\alpha - g_\varepsilon\|^2{}_Y + c_1\,\varepsilon^2\,\|f_\alpha\|_X^2 \quad .$$

Combining the inequalities (5.14), (5.15) we get

$$\|f_\alpha\|^2{}_X \leqslant (1 + c_2\|f^+\|_X^2)/c_1 \tag{5.16}$$

$$\|Af_\alpha - g_\varepsilon\|_Y^2 \leqslant \varepsilon^2\,(1+c_2\,\|f^+\|_X^2) \quad . \tag{5.17}$$

Therefore Af_α strongly converges to g and f_α has a weak limit. Since $f_\alpha \in N(A)^\perp$, the weak limit must be f^+. □

Condition (5.11) is satisfied by $\alpha = c\,\varepsilon$, where c is an arbitrary constant. Therefore this choice of the regularization parameter provides an approximate solution which always converges to f^+ when $\varepsilon \longrightarrow 0$. The convergence, however, can be arbitrarily slow, so that, for a finite value of ε, the approximation can be quite poor. Notice also that condition (5.12) corresponds to the choice discussed in Sect. 4.2. As shown in Theorem 4.3, by an appropriate choice of the constant, the result of weak convergence can be improved.

The previous discussion gives some indications about the relation between regularization parameter and error on the data, but it does not provide a precise recipe for the determination of α in a practical problem. Two methods have already been discussed in the previous Section - Theorem 4.2 and Theorem 4.3: both require a prescribed bound on the norm of the solution and, if the bound is not precise, strong convergence is not assured. In a practical problem, however, it may be more reasonable to give an upper bound on the error rather than an upper bound on the solution. If ε is such an upper bound, then a reasonable criterion for the choice of α consists in requiring that the data function computed using the approximate solution $f_\alpha = R_\alpha g_\varepsilon$, i.e. Af_α, is consistent with the experimental data g_ε within the error ε. The norm of their difference is the discrepancy function $\rho(\alpha)$ - see eq.(4.8) - and therefore the previous criterion, known also as the discrepancy principle, consists of determining the value of α such that

$$\rho(\alpha) = \|Af_\alpha - g_\varepsilon\|_Y = \varepsilon \quad . \tag{5.18}$$

From the results of Lemma 4.2 it follows that if g_ε satisfies the conditions

$$\|Q\, g_\varepsilon\|_Y < \varepsilon < \|g_\varepsilon\|_Y \tag{5.19}$$

then there exists a unique value of α which solves equation (5.18). Notice that the conditions (5.19) are quite reasonable. The component Qg_ε, indeed, can only be an effect of the noise and therefore it must be smaller than the estimate of the error. On the other hand the second inequality implies that the data function is greater than the noise.

It is interesting to reformulate the previous problem as the dual of the problem considered in Theorem 4.2 [27]. We also get a result on the convergence of the approximate solution when $\varepsilon \longrightarrow 0$.

Theorem 5.2 - For any g_ε satisfying the conditions (5.19) there exists a unique solution $\bar{f}_\varepsilon$ of the problem

$$\|\bar{f}_\varepsilon\|_X = \inf\left\{ \|f\|_X \mid \|Af - g_\varepsilon\|_Y \leq \varepsilon \right\} \tag{5.20}$$

which is the unique $f_\alpha = R_\alpha g_\varepsilon$ solving equation (5.18). Furthermore, when $\varepsilon \longrightarrow 0$, i.e. $g_\varepsilon \longrightarrow g \in R(A)$, $\bar{f}_\varepsilon$ strongly converges to f^+, the generalized solution associated with g.

Proof: The set $J_\varepsilon(g_\varepsilon)$, eq.(4.1), is closed and convex and $0 \notin J_\varepsilon(g_\varepsilon)$

since $\|g_\varepsilon\|_Y > \varepsilon$. Therefore there exists a unique element $\bar{f}_\varepsilon \neq 0$ of minimal norm, which lies on the boundary of $J_\varepsilon(g_\varepsilon)$, i.e. satisfies the condition $\|A\bar{f}_\varepsilon - g_\varepsilon\|_Y = \varepsilon$. It follows that $\bar{f}_\varepsilon$ can be determined using the method of Lagrange multipliers: minimize the functional (4.6) for any $\alpha > 0$ and look for the minimum point $f_\alpha = R_\alpha g_\varepsilon$ which solves the equation $\rho(\alpha) = \varepsilon$. From the properties of $\rho(\alpha)$ stated in Lemma 4.2 and the conditions (5.19) on g_ε , it follows that there exists a unique value of α solving the previous equation.

As concerns the convergence, let $\alpha = \alpha(\varepsilon)$ be the value of the regularization parameter obtained with the method described above. Then we have

$$\varepsilon^2 + \alpha\|\bar{f}_\varepsilon\|_X^2 = \|A\bar{f}_\varepsilon - g_\varepsilon\|_Y^2 + \alpha\|\bar{f}_\varepsilon\|_X^2 \leqslant \tag{5.21}$$

$$\leqslant \|Af^+ - g_\varepsilon\|_Y^2 + \alpha\|f^+\|_X^2 \leqslant \varepsilon^2 + \alpha\|f^+\|_X^2$$

where the minimum property of $\bar{f}_\varepsilon$ has been used, and therefore

$$\|\bar{f}_\varepsilon\|_X \leqslant \|f^+\|_X \quad . \tag{5.22}$$

Since $A\bar{f}_\varepsilon$ strongly converges to g, it follows, as in the previous Theorems, that $\bar{f}_\varepsilon$ weakly converges to f^+. Then, from the weak lower semicontinuity of the norm and inequality (5.22) we have

$$\|f^+\|_X \leqslant \text{lim.inf.}\ \|\bar{f}_\varepsilon\|_X \leqslant \text{lim.sup.}\|\bar{f}_\varepsilon\|_X \leqslant \|f^+\|_X \tag{5.23}$$

and therefore

$$\lim \|\bar{f}_\varepsilon\|_X = \|f^+\|_X \quad . \tag{5.24}$$

This result, combined with the weak convergence, implies the strong convergence. □

We can summarize the methods for the choice of the regularization parameter, investigated in the previous Section and in this Section, as follows: the method of Sect.4.1 uses a prescribed bound on the solution; the method of Sect.4.2 uses a prescribed bound both on the solution and on the error; the method of this section uses only a prescribed bound on the error. It is interesting to notice that the last one, which is the most reasonable in practice, is also the unique one of the three methods which always assures strong convergence to the exact solution or generalized solution when the error on the data tends to zero.

5.2. Spectral windows

The regularizer (5.8) can be written in the following form

$$R_\alpha = U_\alpha(A^*A)\, A^* \tag{5.25}$$

where

$$U_\alpha(\lambda) = (\lambda + \alpha)^{-1};\ \lambda \geq 0 \quad . \tag{5.26}$$

This remark suggests a way for defining a wide class of regularization algorithms which can be applied whenever the spectral representation of the operator A^*A is known [4,21].

Consider a family of real valued, piecewise continuous functions $\{U_\alpha\}_{\alpha>0}$, defined on the interval $[0,\ \|A\|^2]$ and assume that they satisfy the following conditions:

i) for each $\alpha > 0$, there exists a constant c_α such that

$$|U_\alpha(\lambda)| \leq c_\alpha\ ;\ \forall \lambda \in [0,\ \|A\|^2]\ ;$$

ii) for each $\alpha > 0$

$$0 \leq \lambda U_\alpha(\lambda) \leq 1\ ;\ \forall \lambda \in [0,\ \|A\|^2];$$

iii) $\lim_{\alpha \downarrow 0} \lambda U_\alpha(\lambda) = 1\ ;\ \forall \lambda \in (0,\ \|A\|^2].$

Then the family of operators R_α, defined by equation (5.25) is a regularization algorithm for A^+. This result derives from the following remarks. Thanks to condition i), for each $\alpha > 0$, the operator R_α is bounded. Furthermore the following relation holds true

$$U_\alpha(A^*A)\, A^* = A^*\, U_\alpha(AA^*) \tag{5.27}$$

since it is true for polynomials and therefore it is also true for continuous functions. This relation implies that, for any $\alpha > 0$ and any $g \in Y$

$$f_\alpha = R_\alpha g \in N(A)^\perp \quad . \tag{5..28}$$

Finally, from the spectral representation of A^*A, from conditions ii), iii) and the dominated convergence theorem, property (5.2) can be derived.

The function $U_\alpha(\lambda)$ defined in eq.(5.26) satisfies the previous conditions. Another important example is given by

$$U_\alpha(\lambda) = 0 \ , \quad 0 \leq \lambda \leq \alpha \ ; \qquad U_\alpha(\lambda) = \lambda^{-1}, \lambda > \alpha \quad . \tag{5.29}$$

For a regularization algorithm of the type (5.25), the operator T_α, equation (5.3), is

$$T_\alpha = W_\alpha(A^*A) \tag{5.30}$$

where $W_\alpha(\lambda) = \lambda\, U_\alpha(\lambda)$. The function $W_\alpha(\lambda)$ can be called a <u>spectral window</u>. The justification of this name derives from the example (5.29) since in such a case the function $W_\alpha(\lambda)$ is zero in a neighbourhood of $= 0$ and is unity elsewhere.

In the case of a compact operator, the regularizer (5.25) can be expressed in terms of the singular system of the operator

$$R_\alpha g = \sum_{k=0}^{+\infty} \lambda_k^{-1/2}\, W_\alpha(\lambda_k)\, (g, v_k)_Y\, u_k \quad . \tag{5.31}$$

In particular, in the case (5.29) we have

$$R_\alpha g = \sum_{\sqrt{\lambda_k} > \alpha} \lambda_k^{-1/2}\, (g, v_k)_Y\, u_k \tag{5.32}$$

and this is the well-known method of truncated singular function expansions [36,21,6].

Another important case where the method of spectral windows can be easily applied is the inversion of convolution operators. We will explicitly consider the case

$$(Af)(x) = \int_{-\infty}^{+\infty} K(x-y)f(y)\, dy \tag{5.33}$$

which can be treated by means of the Fourier transform. However, if the Fourier transform is replaced by the Mellin transform, the same results apply also to the inversion of operators of the following type [54]

$$(Af)(x) = \int_{0}^{+\infty} K(\frac{x}{y})f(y)\, \frac{dy}{y} \tag{5.34}$$

$$(Af)(x) = \int_{0}^{+\infty} K(xy)f(y)dy \quad . \tag{5.35}$$

Notice that Laplace transform is just a particular case of an operator of the form (5.35).

Assume that the function $K(x)$ in equation (5.33) is integrable, so that its Fourier transform $\hat{K}(\xi)$ is bounded, continuous and tends to zero when $|\xi| \longrightarrow +\infty$. Furthermore, assume, for simplicity, that $\hat{K}(\xi)$ has no zeros. Then the operator (5.33) has an inverse given by

$$(A^{-1}g)(x) = \frac{1}{2\pi} \int_{-\infty}^{+\infty} [\hat{K}(\xi)]^{-1}\, \hat{g}(\xi)\, e^{ix\xi}\, d\xi \tag{5.36}$$

whose domain is the set of all the functions g such that $(\hat{g}/\hat{K}) \in L^2(\mathbb{R})$.

A family of window functions is a family of piecewise continuous functions $\{\hat{W}_\alpha\}_{\alpha>0}$ satisfying the conditions:

i) $0 \leqslant \hat{W}_\alpha(\xi) \leqslant 1$;

ii) $\lim\limits_{\alpha \downarrow 0} \hat{W}_\alpha(\xi) = 1$, a.e.;

iii) for any $\alpha > 0$, there exists a constant c_α such that

$|\hat{W}_\alpha(\xi) \,/\, \hat{K}(\xi)| \leqslant c_\alpha < +\infty$, a.e.

Then it is easy to verify that the family of operators

$$(R_\alpha g)(x) = \frac{1}{2\pi} \int_{-\infty}^{+\infty} \hat{W}_\alpha(\xi) \left[\hat{K}(\xi)\right]^{-1} \hat{g}(\xi)\, e^{ix\xi} d\xi \tag{5.37}$$

is a regularization algorithm for the inversion of A. The corresponding operators T_α, equation (5.3), are given by

$$(T_\alpha f)(x) = \int_{-\infty}^{+\infty} W_\alpha(x - y)\, f(y)\, dy \tag{5.38}$$

where W_α in the inverse Fourier transform of $\hat{W}_\alpha$.

Regularization by means of truncation of the Fourier integral corresponds to taking

$$\hat{W}_\alpha(\xi) = \chi_\alpha(\xi) \tag{5.39}$$

where $\chi_\alpha(\xi)$ is the characteristic function of the interval $[-1/\alpha, 1/\alpha]$. Analogously the use of the triangular window

$$\hat{W}_\alpha(\xi) = (1 - \alpha|\xi|)\, \chi_\alpha(\xi) \tag{5.40}$$

is equivalent to approximating the Fourier integral in the sense of (C,1) - summability [54]. Notice that, in the case (5.39)

$$W_\alpha(x) = (\pi\alpha)^{-1} \left[\sin(x/\alpha)\right] / (x/\alpha) \tag{5.41}$$

while in the case (5.40)

$$W_\alpha(x) = (2\pi\alpha)^{-1} \left[\sin(x/2\alpha)\right]^2 / (x/2\alpha)^2 \tag{5.42}$$

and that the last function is positive. Therefore, in the absence of errors on the data, the triangular window provides positive approximations of positive functions.

Another window, with the same property, which is often used in the approximate computation of the derivative of a function (see Sect.2), is the gaussian window.

$$\hat{W}_{\alpha}(\xi) = \exp(-\alpha\xi^2) \tag{5.43}$$

whose inverse Fourier transform is

$$W_{\alpha}(x) = (2\pi\alpha)^{-1/2} \exp(-x^2/2\alpha) \quad . \tag{5.44}$$

For all the spectral windows considered in this section, the discrepancy principle provides a unique solution to the problem of choosing the regularization parameter. This result derives from the remark that the spectral windows we have considered, $W_{\alpha}(\lambda)$ or $\hat{W}_{\alpha}(\xi)$, are decreasing functions of α for any λ or ξ and that they tend to zero when α tends to infinity. These properties imply that the discrepancy function

$$\rho(\alpha) = \| AR_{\alpha}g - g \|_Y = \| W_{\alpha}(AA^*)\, g - g \|_Y \tag{5.45}$$

is an increasing function with values in the interval $(\|Qg\|_Y, \|g\|_Y)$. Therefore, if the noisy data g_{ε} satisfy the condition (5.19), there exists a unique solution of the equation $\rho(\alpha) = \varepsilon$. It is also obvious that, when $\varepsilon \longrightarrow 0$, such a value of α, $\alpha(\varepsilon)$, tends to zero. However the question of the convergence of $R_{\alpha}g$ to f^+ in the general case seems to be open.

5.3. Iterative methods

Some well-known iterative methods for the computation of least squares solutions, such as the methods of successive approximations, steepest descent and conjugate gradient, have interesting regularizing properties. It has been proved indeed [30-32] that each of them provides a regularization algorithm, the regularization parameter being related to the number of iterations.

These methods apply to the problem of determining the generalized solution, which is the solution of minimal norm of equation (3.3). Denote by f_n the approximation provided by the n-th iteration and put

$$r_n = A^*A\, f_n - A^*g \quad . \tag{5.46}$$

Then the above mentioned algorithms can be formulated as follows:

a) Successive approximations

$$f_0 = 0 \quad , \quad f_{n+1} = f_n - \tau r_n \tag{5.47}$$

where τ is a fixed constant (relaxation parameter) satisfying the condition

$$0 < \tau < 2/\|A\|^2 \quad . \tag{5.48}$$

b) Steepest descent

$$f_0 = 0 \quad , \quad f_{n+1} = f_n - \tau_n r_n \tag{5.49}$$

where

$$\tau_n = \|r_n\|_X^2 / \|Ar_n\|_Y^2 \quad . \tag{5.50}$$

c) Conjugate gradient

$$f_0 = 0 \quad , \quad f_{n+1} = f_n - \tau_n p_n \tag{5.51}$$

where

$$p_0 = r_0 = -A^*g \quad , \quad p_n = r_n + \sigma_{n-1} p_{n-1} \tag{5.52}$$

$$\tau_n = (r_n, p_n)_X / \|Ap_n\|_Y^2 \; , \; \sigma_{n-1} = -(Ar_n, Ap_{n-1})_Y / \|Ap_{n-1}\|_Y^2 \quad . \tag{5.53}$$

It has been proved [30-32] that, each of these algorithms provides a sequence $\{f_n\}$ which strongly converges to f^+ when $n \longrightarrow \infty$ provided that $Pg \in R(A)$ in the case a), $Pg \in R(AA^*)$ in the case b) and $Pg \in R(AA^*A)$ in the case c). Therefore if we put $f_n = R_n g$, then $\{R_n\}$ is a regularization algorithm, according to the general definition, the role of the regularization parameter being played by the number of iterations. It is easy to verify indeed that $R_n : Y \longrightarrow X$ is continuous. Notice also that in the case a) we have a linear algorithm while R_n is nonlinear in the other cases.

The method of successive approximations is a particular case of the spectral window method. From eqs.(5.47) and (5.46) it can be derived that

$$f_n = R_n g = \tau \sum_{k=0}^{n-1} (I - \tau A^*A)^k A^*g \tag{5.54}$$

and therefore R_n has the form (5.25) with

$$U_n(\lambda) = \tau \sum_{k=0}^{n-1} (1 - \tau\lambda)^k = \frac{1}{\lambda}\left[1 - (1 - \tau\lambda)^n\right] \tag{5.55}$$

while the corresponding spectral window is

$$W_n(\lambda) = 1 - (1 - \tau\lambda)^n \quad . \tag{5.56}$$

The discrepancy principle can be used also in the case of the iterative methods for choosing an optimum number of iterations since the discrepancy function is a decreasing function of n. Therefore one can stop the iteration when the value of the discrepancy function is just above or just below the threshold value ε . In this way it is easy to test that the convergence of the conjugate gradient method is much faster than the convergence of the two other methods. But, for a given value of ε , the three methods provide different approximations and which of them is better is a question which must be investigated in any specific practical problem. Also the convergence of the approximation provided by the discrepancy principle when $\varepsilon \longrightarrow 0$ seems to be an open question.

5.4. Regularization algorithms for C-generalized inverses

In the cases where the C-generalized inverse of an operator A is not continuous, one needs regularization algorithms for its approximation. As it will be shown, this result can be obtained by minimizing the functional

$$\Phi_\alpha[f] = \|Af - g\|_Y^2 + \alpha \|Cf\|_Z^2 \tag{5.57}$$

for any $\alpha > 0$. As a byproduct one also obtains a method for approximating the solution of the equation $Af = g$, when A has an inverse and $g \in AD(C)$.

Let us assume that the constraint operator C satisfies the conditions i)-iii) of Sect.3.4. Then, for any $g \in Y$, there exists a unique function f_α which minimizes the functional (5.57). In order to prove this result, let us introduce, as in Sect.3.4, the space $W = Y \oplus Z$ and the operator $B_\alpha : X \longrightarrow W$ defined as follows: $D(B_\alpha) = D(C)$, $B_\alpha f = \{Af, \alpha Cf\}$. Then the functional (5.57) can be written in the form

$$\Phi_\alpha[f] = \|B_\alpha f - v\|_W^2 \tag{5.58}$$

where $v = \{g, 0\}$. Since $R(B_\alpha)$ is closed - see the proof in the case of the operator B, eq.(3.19) - there exists a unique $\bar{v}_\alpha \in R(B_\alpha)$ which is closest to v. Finally the equation $B_\alpha f = \bar{v}_\alpha$ has a unique solution f_α because B_α has an inverse. Furthermore f_α can be obtained by solving the Euler equation

$$(A^*A + \alpha C^*C)\, f_\alpha = A^*g \quad . \tag{5.59}$$

Since $R(B_\alpha)$ is closed, B_α^{-1} is continuous and therefore the operator $R_\alpha : Y \longrightarrow X$ defined by $f_\alpha = R_\alpha g$ is also continuous. We can prove now that $\{R_\alpha\}_{\alpha>0}$ is a regularization algorithm for the C-generalized inverse A_C^+ of A.

<u>Theorem 5.3</u> - If $Pg \in A\, D(C)$ and $\alpha \longrightarrow 0$, then f_α strongly converges to f_C^+, the C-generalized solution associated with g.

<u>Proof</u>: From the minimum property of f_α we have

$$\|Af_\alpha - g\|_Y^2 + \alpha \|Cf_\alpha\|_Z^2 \leq \tag{5.60}$$

$$\leq \|Af_C^+ - g\|_Y^2 + \alpha \|Cf_C^+\|_Z^2 = \mu_g^2 + \alpha \nu_g^2$$

μ_g and ν_g being defined in eqs.(3.21),(3.22). It follows that, for any $\alpha\in[0,1]$, $\|Af_\alpha - g\|_Y$ is bounded by a constant independent of α

$$\|Af_\alpha - g\|_Y \leqslant [\mu_g^2 + \nu_g^2]^{1/2} \quad . \tag{5.61}$$

Furthermore, thanks to the minimum property of f_C^+

$$\mu_g = \|Af_C^+ - g\|_Y \leqslant \|Af_\alpha - g\|_Y \tag{5.62}$$

and this inequality, combined with (5.60), gives

$$\|Cf_\alpha\|_Z \leqslant \nu_g \quad . \tag{5.63}$$

Finally, from the condition (3.18) it follows that $\|f_\alpha\|_X$ is also bounded by a constant independent of α. Therefore it is possible to find a sequence of positive numbers α_n with $\alpha_n \longrightarrow 0$ and such that, if we put $f_n = f_{\alpha_n}$, then the three sequences $\{Af_n\}$, $\{Cf_n\}$, $\{f_n\}$ are weakly convergent. Let f be the limit of f_n. Since A is linear and continuous, $Af_n \rightharpoonup Af$. Analogously, since C is weakly closed (C is closed and R(C) = Z), it follows $f\in D(C)$ and $Cf_n \rightharpoonup Cf$.

Now, from the inequalities (5.62), (5.60) and the weak lower semicontinuity of the norm, we have

$$\mu_g \leqslant \|Af - g\|_Y \leqslant \text{lim.inf.}\|Af_n - g\|_Y \leqslant \tag{5.64}$$

$$\leqslant \text{lim.sup.}\|Af_n - g\|_Y \leqslant \lim\left[\mu_g^2 + \alpha_n \nu_g^2\right]^{1/2} = \mu_g$$

and this implies $\|Af - g\|_Y = \mu_g$, i.e. $f \in S_g \cap D(C)$, $\|Cf\|_Z \geqslant \nu_g$. Then, combining this result with the inequality (5.63)

$$\nu_g \leqslant \|Cf\|_Z \leqslant \text{lim.inf.}\|Cf_n\|_Z \leqslant \text{lim.sup.}\|Cf_n\|_Z \leqslant \nu_g \tag{5.65}$$

and therefore $\|Cf\|_Z = \nu_g$. It follows that $f = f_C^+$ and that Af_n and Cf_n strongly converge to Af_C^+ and Cf_C^+ respectively. Using again the condition (3.18), we conclude that also f_n strongly converges to f_C^+.□

As concerns the choice of the regularization parameter, it is possible to prove [38] that the discrepancy principle provides a unique value of α and that, when the error tends to zero, the corresponding approximate solution strongly converges to the C-generalized solution of the problem. The proof requires some rather natural extensions of Lemma 4.2 and Theorem 5.2.

6. CONVERGENCE RATES AND STABILITY ESTIMATES

In the previous Section we have given several results concerning the convergence of various regularization algorithms in the case where the error on the data tends to zero. No result, however, has been given on the rate of convergence since such results require further specification of the set of the solutions.

Estimates of the convergence rates can be interesting both in theory and in practice since they provide an indication of the accuracy of the approximate solution and also of the possibility of improving the approximation by reducing the error on the data.

For simplicity we will mainly consider the case where the operator A has an inverse A^{-1}. Several results, however, can be easily extended to the case of the generalized inverse A^{+}, just by restricting A to $N(A)^{\perp}$.

We first define a convergence rate of the regularization algorithm $\{R_\alpha\}_{\alpha>0}$ in the case of exact data associated with functions f of a prescribed set H

$$\omega_H(\alpha) = \sup\{\|R_\alpha Af - f\|_X \mid f \in H\} \quad . \tag{6.1}$$

An extimate of $\omega_H(\alpha)$, combined with inequality (5.5), can be used in order to find a choice of the regularization parameter which is optimum for the set H [21].

In the case of noisy data corresponding, within an error level ε, to solutions in the set H it is convenient to introduce the following modulus of convergence [17] of the regularization algorithm $\{R_\alpha\}_{\alpha>0}$

$$\sigma_H(\varepsilon,\alpha) = \sup\{\|R_\alpha g - f\|_X \mid f \in H,\ \|Af - g\|_Y \leq \varepsilon\} \quad . \tag{6.2}$$

The inequality (5.5) implies that

$$\sigma_H(\varepsilon,\alpha) \leq \omega_H(\alpha) + \varepsilon N(\alpha) \quad . \tag{6.3}$$

Finally we introduce the modulus of continuity of the operator A^{-1} when restricted to AH :

$$\mu_H(\varepsilon) = \sup\{\|f\|_X \mid f \in H, \|Af\|_Y \leq \varepsilon\} \quad . \tag{6.4}$$

If the set H contains a neighbourhood of 0, then $\mu_H(\varepsilon)$ is a continuous, increasing function of ε. Furthermore, if the set H is compact, then Lemma 4.1 implies that $\mu_H(\varepsilon) \longrightarrow 0$. We point out that the compactness of H is only a sufficient condition: it is easy to find examples (and we will give one of them in the following) of weakly compact sets H such that $\mu_H(\varepsilon) \longrightarrow 0$. Any upper bound for $\mu_H(\varepsilon)$ will be called a stability estimate [36].

The next result [17] enlightens the relationship between the modulus of convergence and the modulus of continuity.

Theorem 6.1 - If the set H contains a neighbourhood of 0, then for any linear regularization algorithm $\{R_\alpha\}_{\alpha>0}$ and any α, the following inequality holds true:

$$\mu_H(\varepsilon) \leq \sigma_H(\varepsilon,\alpha) \quad . \tag{6.5}$$

Proof: Since H contains a neighbourhood of 0, there exists $u \in H$ such that $\|Au\|_Y \leq \varepsilon$. Then the pair $\{f = u,\ g = 0\}$ satisfies the conditions $f \in H$, $\|Af - g\|_Y \leq \varepsilon$. Furthermore, thanks to the linearity of R_α, we have $\|R_\alpha g - f\|_X = \|f\|_X$ and therefore: $\sigma_H(\varepsilon,\alpha) \geq \|f\|_X$. But this implies $\sigma_H(\varepsilon,\alpha) \geq \sup\{\|f\|_X \mid f \in H, \|Af\|_Y \leq \varepsilon\}$, which is precisely the inequality (6.5). □

The relevance of the result stated in the previous Theorem is obvious: given a set H of solutions, no regularization algorithm and no choice of the regularization parameter can provide a modulus of convergence which tends to zero more rapidly than the modulus of continuity and therefore the last one is the best possible convergence rate for the approximation of elements of the set H using noisy data.

It is interesting to notice that in some cases the H-constrained least squares solutions introduced in Sect.4.1 just provide this optimum convergence rate [26].

Theorem 6.2 - Let the set H satisfy the conditions of Theorem 4.1 and Theorem 6.1 and furthermore let H be symmetric with respect to 0. Then, if f, f' are the H-constrained least squares solutions associated with g, g' and $\|g - g'\|_Y \leq \varepsilon$, for sufficiently small ε

$$\|\tilde{f} - \tilde{f}'\|_X \leq \mu_H(\varepsilon) \quad . \tag{6.6}$$

Proof: Notice that $\|A(\tilde{f} - \tilde{f}')\|_Y = \|\tilde{g} - \tilde{g}'\|_Y \leq \|g - g'\|_Y \leq \varepsilon$, where $\tilde{g}$, $\tilde{g}'$ are the convex projections of g, g' onto AH. Then

$$\|\tilde{f} - \tilde{f}'\|_X \leq \sup\{\|f - f'\|_X \mid f, f' \in H, \|A(f - f')\|_Y \leq \varepsilon\} \quad . \tag{6.7}$$

Since $\mu_H(\varepsilon) \longrightarrow 0$ and H is symmetric with respect to 0, for sufficiently small ε one can take f' = 0 in (6.7) and (6.6) is proved.□

If H is a level set of a functional of the type (3.14), as given by eq.(4.16), then the following relation holds true

$$\mu_H(\varepsilon) = E\mu_C(\varepsilon/E) \tag{6.8}$$

where

$$\mu_C(\varepsilon) = \sup\{\|f\|_X \mid \|Cf\|_Z \leq 1, \|Af\|_Y \leq \varepsilon\} \quad . \tag{6.9}$$

This modulus of continuity, when it tends to zero, gives the rate of convergence of the least squares solutions introduced in Sect.4.2 [36].

Theorem 6.3 - If the compatibility condition (4.24) is satisfied and K is the set defined in equation (4.2) then, for any $f \in K$

$$\|\tilde{f}_0 - f\|_X < \sqrt{2}\,\mu_C(\varepsilon) \tag{6.10}$$

where $\tilde{f}_0$ is the minimum point of the functional (4.21).

Proof: When (4.24) is satisfied, $\tilde{f}_0 \in K$. Then notice that the functional (4.21) can be written in the form (5.58) with $\alpha = \varepsilon^2$ and therefore $B_\alpha \tilde{f}_0$ is the orthogonal projection of $v = \{g, 0\}$ onto $R(B_\alpha) \subset W = Y \oplus Z$. Since $(B_\alpha \tilde{f}_0 - v) \in R(B_\alpha)^\perp$, it follows that

$$\|B_\alpha(f - \tilde{f}_0)\|_W^2 + \|B_\alpha \tilde{f}_0 - v\|_W^2 = \|B_\alpha f - v\|_W^2 =$$

$$= \|Af - g\|_Y^2 + \varepsilon^2 \|Cf\|_Z^2 \leq 2\varepsilon^2 \quad , \quad (\alpha = \varepsilon^2) \tag{6.11}$$

where, in the last inequality, the fact that $f \in K$ has been used. This implies

$$\|B_\alpha(f - \tilde{f}_0)\|_W^2 = \|A(f - \tilde{f}_0)\|_Y^2 + \varepsilon^2 \|C(f - \tilde{f}_0)\|_Z^2 \leq 2\,\varepsilon^2 \tag{6.12}$$

and therefore

$$\|A(f - \tilde{f}_0)\|_Y \leq \sqrt{2}\,\varepsilon, \quad \|C(f - \tilde{f}_0)\|_Z \leq \sqrt{2} \quad . \tag{6.13}$$

These inequalities and the relation (6.8) prove the Theorem. □

The inclusions (4.25), in the case $g = 0$, provide a stability estimate for $\mu_C(\varepsilon)$ or, more precisely, they provide both an upper and a lower bound for $\mu_C(\varepsilon)$. If we put

$$\mu_C^{(0)}(\varepsilon) = \sup \{ \|f\|_X \mid \|Af\|_Y^2 + \varepsilon^2 \|Cf\|_Z^2 \leq \varepsilon^2 \} \tag{6.14}$$

then we have

$$\mu_C^{(0)}(\varepsilon) \leq \mu_C(\varepsilon) \leq \sqrt{2}\,\mu_C^{(0)}(\varepsilon) \tag{6.15}$$

and therefore, except for a factor which is at most $\sqrt{2}$, $\mu_C^{(0)}(\varepsilon)$ is an optimum stability estimate of the modulus of continuity. It is also important to notice that the computation of $\mu_C^{(0)}(\varepsilon)$ can be reduced to the solution of a spectral problem for the operator $A^*A + \varepsilon^2 C^*C$. As we know, this operator is positive definite and therefore its spectrum has a positive lower bound $\gamma^2(\varepsilon)$

$$\gamma^2(\varepsilon) = \inf \{ \|Af\|_Y^2 + \varepsilon^2 \|Cf\|_Z^2 \mid \|f\|_X = 1 \} \quad . \tag{6.16}$$

Then, comparing with (6.14) we have

$$\mu_C^{(0)}(\varepsilon) = \varepsilon/\gamma(\varepsilon). \tag{6.17}$$

At this point it is easy to give an example showing that the compactness of H is only a sufficient condition for strong stability. Consider the case of the convolution operator (5.33) and assume that the kernel K(.) satisfies all the conditions which have been assumed in that case. Introduce a constraint operator C such that the corresponding functional (3.14) is given by

$$\|Cf\|_X^2 = \frac{1}{2\pi}\int_{-\infty}^{+\infty} |\hat{P}(\xi)|^2 |\hat{f}(\xi)|^2 d\xi \tag{6.18}$$

where $\hat{P}(.)$ is, for example, a polynomial of degree n, i.e. C is a differential operator of order n with constant coefficients. Assume also that $|\hat{P}(\xi)|$ has a lower bound so that C has a continuous inverse C^{-1} with $D(C^{-1}) = X$. It follows that the level sets of $\|Cf\|_X$ are only weakly compact since C^{-1} is not compact. Then, from (6.16) we have

$$\gamma(\varepsilon) = \inf_{\xi} \left\{ |\hat{K}(\xi)|^2 + \varepsilon^2 |\hat{P}(\xi)|^2 \right\}^{1/2} \quad . \tag{6.19}$$

Assume, for simplicity, that $|\hat{K}(\xi)|$ and $|\hat{P}(\xi)|$ are even functions and that there exists $\xi_0>0$ such that, when $|\xi| > \xi_0$, both $|\hat{K}(\xi)|$ and $|\hat{P}(\xi)|$ are monotone functions. Then $|\hat{K}(\xi)|$ is a decreasing function which tends to zero when $|\xi| \longrightarrow \infty$ and $|\hat{P}(\xi)|$ is an increasing function which tends to infinity. Under these circumstances, when ε is sufficiently small, the lower bound $\gamma(\varepsilon)$ is reached at a point $\xi(\varepsilon) > \xi_0$ and $\xi(\varepsilon) \rightarrow +\infty$ when $\varepsilon \rightarrow 0$. Since $\gamma(\varepsilon) \geqslant \varepsilon|\hat{P}[\xi(\varepsilon)]|$, from eq.(6.17) we get

$$\mu_C^{(0)}(\varepsilon) \leqslant |\hat{P}[\xi(\varepsilon)]|^{-1} \longrightarrow 0, \varepsilon \longrightarrow 0 \quad . \tag{6.20}$$

Therefore we have strong convergence of the H-constrained least squares solutions even if the set H is not compact. More precise estimates of the behaviour at $\varepsilon=0$ of $\mu_C^{(0)}(\varepsilon)$ can be obtained if one has more informations about the behaviour at infinity of the functions $\hat{K}(\xi)$

and $\hat{P}(\xi)$. If, for example, $|\hat{K}(\xi)| = O(|\xi|^{-p})$, $|\xi| \longrightarrow +\infty$ - this roughly means that the kernel has derivatives up to the order p - and $|\hat{P}(\xi)| = O(|\xi|^{q})$, i.e. we look for solutions with derivatives up to the order q, then

$$\mu_C^{(0)}(\varepsilon) = O(\varepsilon^{\nu}), \ \nu = q/(p+q) \quad . \tag{6.21}$$

On the other hand, if $|\hat{K}(\xi)| = O[\exp(-a|\xi|)]$, i.e. the kernel is analytic, and $|\hat{P}(\xi)|$ satisfies the previous condition, then

$$\mu_C^{(0)}(\varepsilon) = O(|\ln \varepsilon|^{-q}) \quad . \tag{6.22}$$

But, if we take $\hat{P}(\xi)$ such that $|\hat{P}(\xi)| = O[\exp(-b|\xi|)]$, i.e. we look for solutions which are also analytic, then we get an estimate of the type (6.21), with $\nu = b/(a+b)$.

We conclude that, if one looks for solutions whose regularity properties are analogous to the regularity properties of the kernel, then the restored continuity is of the Hölder type. In such a case the ill-posed problem can be called <u>well-behaved</u>, using a terminology introduced by John in a fundamental paper where stability estimates are derived for several ill-posed problems for partial differential equations [29] . On the same subject see also [35]. On the other hand, when the kernel is much more smooth than the solution, then the problem is not well-behaved. This is the case of logarithmic continuity (6.22): in such a case any effort to reduce the error on the data is irrelevant since it does not produce a significant reduction of the error on the solution. Similar results can be derived in the case of compact integral operators using the asymptotic behaviour of the singular values which, as follows from results of Hille and Tamarkin [24] , is related to the regularity properties of the kernel.

Finally we give a few words about stability estimates in the case of weak continuity. In such a case, for each given linear continuous functional of the solution

$$F_{\phi}(f) = (f, \phi)_X \tag{6.23}$$

one can introduce a convergence rate and a modulus of convergence for a given regularization algorithm $\{R_\alpha\}_{\alpha>0}$ and also a modulus of continuity

$$\mu_H(\varepsilon;\phi) = \sup\{\,|(f,\phi)_X| \mid f\in H,\ \|Af\|_Y \leq \varepsilon\}\ . \tag{6.24}$$

The extension to the present case of Theorem 6.1 and of the inequalities (6.15) is straightforward. Equation (6.17) is now replaced by [36]

$$\mu_C^{(0)}(\varepsilon;\phi) = \varepsilon\|(A^*A+\varepsilon^2C^*C)^{-1/2}\phi\|_X\ . \tag{6.25}$$

It is easy to prove that $\mu_C^{(0)}(\varepsilon;\phi)\longrightarrow 0,\ \varepsilon\longrightarrow 0$ for any ϕ, whenever the operator C has a bounded inverse.

It is important to notice that, for any linear inverse problem there exists a class of linear continuous functionals which can be estimated directly from the data so that no use of regularization algorithms is required. This is the case when $\phi\in R(A^*)$ so that we can write $\phi = A^*u$ with $u\in N(A^*)^{\perp} = \overline{R(A)}$ and therefore $F_\phi(f) = (Af,u)_Y$. Then the estimated value of this functional is just $(g,u)_Y = (Pg,u)_Y$ where g is the experimental data function. If the norm of the error on g is of the order of ε, then the error in the estimation of the functional is of the order of $\varepsilon\|u\|_Y$.

7. INVERSE PROBLEMS WITH DISCRETE DATA

A linear inverse problem with discrete data can be formulated as follows [7]: given a set of N linear continuous functionals on the object space X, say $F_n(f) = (f,\phi_n)_X$, n=1,...,N, and given a set $\{g_n\}_{n=1}^N$ of values of these functionals, find a functions $f\in X$ such that

$$(f,\phi_n)_X = g_n\ ;\ n = 1,\ldots,N\ . \tag{7.1}$$

The set of the numbers g_n is the data of the problem and we will denote by $\underline{g}$ the vector with components g_n. The data space is therefore an N-dimensional vector space Y_N and we can assume that Y_N is equipped with a scalar product given, in general, by

$$(\underline{g},\underline{h})_N = \sum_{n,m=1}^{N} w_{nm} g_m h_n \tag{7.2}$$

the choice of the positive definite weight matrix $[W]$ being dictated by physical considerations. In linear regression theory, for instance, $[W]$ is the inverse of the covariance matrix of the noise which contaminates the data g_n.

Define the operator $A_N : X \longrightarrow Y_N$ as the operator which transforms a function $f \in X$ into a vector of Y_N whose n-th component is given by

$$(A_N f)_n = (f, \phi_n)_X \; ; \; n = 1,\ldots,N \quad . \tag{7.3}$$

Then the problem (7.1) takes the form:given $\underline{g} \in Y_N$, find $f \in X$ such that

$$A_N f = \underline{g} \tag{7.4}$$

which has precisely the same structure as equation (1.1).

Examples of problems which can be formulated in this way are the following:

1) problems with intrinsically discrete data such as, for instance: the moment problem, i.e. find a function $f \in L^2(0,1)$, given a finite set of its moments

$$\int_0^1 x^{n-1} f(x)dx = \mu_n ; \; n = 1,\ldots,N \; , \tag{7.5}$$

the determination of a function $f \in L^2(0,\pi)$, given a finite set of its Fourier coefficients

$$\int_0^\pi \sin(nx) f(x)dx = c_n ; \; n = 1,\ldots,N \; , \tag{7.6}$$

and linearized inverse eigenvalue problems such as the problem of the inversion of gross Earth data [1];

2) the interpolation problem, in the case where the interpolating function belongs to a reproducing kernel Hilbert space, and the

related problem of the numerical differentiation of a function specified on a finite set of points;

3) the method of moment discretization (or collocation) for the approximate solution of first kind Fredholm integral equations [40];
4) any problem taking the form (1.1), when the data function g is determined by means of a finite number of detectors which measure weighted integrals of g

$$g_n = \int \psi_n(x)(Af)(x)\, dx \tag{7.7}$$

the function ψ_n being related to the position, size, efficiency etc. of the n-th detector.

The range of the operator A_N, defined in equation (7.3), is obviously closed. More precisely, if the functions ϕ_n are linearly independent, then $R(A_N) = Y_N$. When some of the ϕ_n are not linearly independent, $R(A_N)$ is a subspace of Y_N and therefore equation (7.4) does not have a solution for any $\underline{g} \in Y_N$. In such a case it is necessary to introduce least square solutions as in Sect. 3.1. For the sake of simplicity we will only consider the case of linearly independent ϕ_n.

Let $X_N \subset X$ be the subspace spanned by the ϕ_n. Then the null space $N(A_N)$ is just the orthogonal complement of X_N and the set of the solutions of equation (7.4) is the infinite dimensional affine subspace obtained by a translation of $N(A_N)$ - see (3.5). The generalized solution f^+ of equation (7.4), or solution of minimal norm, is the unique solution orthogonal to $N(A_N)$ and therefore it is the unique solution in X_N. It can be obtained by solving a linear algebraic system. If we put

$$f^+ = \sum_{n=1}^{N} a_n \phi_n \tag{7.8}$$

then, from the equation (7.1) we get

$$\sum_{m=1}^{N} G_{mn}\, a_m = g_n ;\ n = 1,\ldots,N \tag{7.9}$$

where $G_{mn} = (\phi_m, \phi_n)_X$ are the matrix elements of the <u>Gram matrix</u> [G] associated with $\phi_1, \ldots, \phi_N$. This is not singular because the ϕ_n are linerly independent.

One can also introduce C-generalized solutions as in Sect.3.4. The results proved in that Section apply also to the present case. We just notice that, if the constraint operator C has a bounded inverse and if D(C) and D(C*) are dense in X and Z resepctively, so that $(C^*)^{-1} = (C^{-1})^*$, then the C-generalized solution of equation (7.4) is given by

$$f^+_C = \sum_{n=1}^{N} b_n \psi_n \tag{7.10}$$

where the ψ_n are the solutions of the equations

$$C^*C\,\psi_n = \phi_n \;;\; n = 1,\ldots,N \tag{7.11}$$

and the coefficients b_n can be obtained by solving the linear algebraic system

$$\sum_{m=1}^{N} \tilde{G}_{mn}\, b_m = g_n \;;\; n = 1,\ldots,N \tag{7.12}$$

with $\tilde{G}_{mn} = (\psi_m, \phi_n)_X$. When the ϕ_n are linearly independent, the matrix $[\tilde{G}]$ is also nonsingular.

In any inverse problem with discrete data, the problem of determining generalized or C-generalized solutions is always well-posed in the sense of Sect. 2. However, it can be ill-conditioned and, very often, extremely ill-conditioned if the number of data points is large. The measure of the ill-conditioning depends on the choice of the metric of Y_N. In order to investigate this problem, it is convenient to introduce the singular system of the operator A_N.

If the scalar product in Y_N is defined by equation (7.2), then the adjoint operator A_N^*: $Y_N \longrightarrow X$ is given by

$$A_N^*\underline{g} = \sum_{n=1}^{N} \Big(\sum_{m=1}^{N} W_{nm} g_m\Big)\, \phi_n \quad . \tag{7.13}$$

The operator $A_N^* A_N$ is a finite rank operator in X, while $A_N A_N^*$ is a linear operator in the vector space Y_N and the associated (nonsingular) matrix is

$$[A_N A_N^*] = [G^T][W] \; . \tag{7.14}$$

Therefore, in order to determine the singular system $\{\sqrt{\lambda_{N,k}};\, u_{n,k},\, \underline{v}_{n,k}\}_{k=0}^{N-1}$ of A_N, one has just to solve a standard eigenvalue problem for the matrix (7.14). The singular values of A_N are the square roots of the eigenvalues of this matrix and the singular vectors $\underline{v}_{N,k}$ are the corresponding ergenvectors. Then the singular functions $u_{N,k}$ can be computed by means of the second of the equations (3.10) which, explicitly written, gives

$$u_{N,k} = \lambda_{N,k}^{-1/2} \sum_{n=1}^{N} \left(\sum_{m=1}^{N} W_{nm} (\underline{v}_{N,k})_m \right) \phi_n \; . \tag{7.15}$$

In terms of the singular system of A_N, the canonical representation of f^+ is as follows

$$f^+ = \sum_{k=0}^{N-1} \lambda_{N,k}^{-1/2} (\underline{g}, \underline{v}_{N,k})_N \, u_{N,k} \tag{7.16}$$

and the condition number of the problem is given by

$$\mathrm{cond}(A_N) = (\lambda_{N,0} / \lambda_{N,N-1})^{1/2} \; . \tag{7.17}$$

When the ϕ_n are orthonormal and $[W] = [I]$, then $[A_N A_N^*] = [I]$ and therefore the condition number is 1. Such a problem is well-conditioned. It is quite easy, however, to find examples of extremely ill-conditioned problems even in the case of moderate values of N. A classical example is the moment problem (7.5), whose Gram matrix is the well-known Hilbert matrix $[H_N(-1)]$. In the case $[W] = [I]$ the singular values of the problem are just the square roots of the eigenvalues of the Hilbert matrix and therefore one can compute its condition number. From results reported in [18] it follows that for $N = 3,4,5,6$ the condition number is, respectively, 22.9, 124.6, 690.6, 3865 and therefore it grows very rapidly with the number of given moments. The following asymptotic behaviour can also be proved for large N: $\mathrm{cond}(A_N) \cong \exp(1.75\, N)$ [18].

All the regularization algorithms developed for ill-posed problems and reviewed in Sects 4,5, can also be applied to the case of an ill-conditioned inverse problem with discrete data. The extension is obvious and only few remarks are required in order to show that they are computable. For instance, in the case of the Tikhonov regularizer (5.8), it is convenient to write the regularized solution in the following form, deduced from equation (4.10)

$$f_{\alpha} = A_N^*(A_N A_N^* + \alpha I)^{-1} \underline{g} \quad . \tag{7.18}$$

This clearly shows that the computation of f_{α} can be essentially reduced to the inversion of an N x N matrix. Similar remarks apply to any other spectral window and also to the iterative methods.

The various methods for the choice of the regularization parameter, discussed in the previous Sections, apply also to the present case. For inverse problems with discrete data, however, a specific method has been proposed which works essentially in the case of Tikhonov regularizer (7.18) and does not require the knowledge of any upper bound on the error or on the solution. This is the method of <u>cross-validation</u> or its modified form known as <u>generalized cross-validation</u> [14,55]. It has been applied to smoothing problems [14] and also to the moment discretization of first kind Fredholm integral equations [55].

The idea behind cross-validation is to allow the data values themselves to choose the value of the regularization parameter by requiring that a good value of the parameter should predict missing data values. In this way no <u>a priori</u> knowledge about the solution and/or the noise is required.

Let $(A_N f)_n$ be defined as in (7.3) and let $f_{\alpha,k}$ be the minimizer of the functional

$$\Phi_{\alpha,k}[f] = \frac{1}{N} \sum_{n \neq k} |(A_N f)_n - g_n|^2 + \alpha \|f\|_X^2 \quad . \tag{7.19}$$

Then the <u>cross-validation function</u> $V_0(\alpha)$ is defined by

$$V_0(\alpha) = \frac{1}{N} \sum_{k=1}^{N} |(A_N f_{\alpha,k})_k - g_k|^2 \tag{7.20}$$

and the cross-validation method consists of determining the unique value of α which minimizes $V_0(\alpha)$. The computation of the minimum is based on the relation [14]

$$V_0(\alpha) = \frac{1}{N} \sum_{k=1}^{N} |1 - \tilde{A}_{kk}(\alpha)|^{-2} |(A_N f_\alpha)_k - g_k|^2 \qquad (7.21)$$

where f_α is the minimizer of the functional

$$\Phi_\alpha[f] = \frac{1}{N} \sum_{k=1}^{N} |(A_N f)_k - g_k|^2 + \alpha \|f\|_X^2 \qquad (7.22)$$

and $\tilde{A}_{kk}(\alpha)$ is the kk-entry of the NxN matrix

$$[\tilde{A}(\alpha)] = [A_N A_N^*][A_N A_N^* + \alpha I]^{-1} \quad . \qquad (7.23)$$

Notice that $[A_N A_N^*]$ is essentially the Gram matrix of the functions ϕ_n, since, in this formulation, $[W] = N^{-1}[I]$.

It has been shown [14] that, from the point of view of minimizing the predictive mean square error, the minimization of $V_0(\alpha)$ must be replaced by the minimization of the <u>generalized cross-validation function</u>, defined by

$$V(\alpha) = \left(\frac{1}{N} \operatorname{Tr}[I - \tilde{A}(\alpha)]\right)^{-2} \left(\frac{1}{N} \|[I - \tilde{A}(\alpha)]\underline{g}\|^2\right) \qquad (7.24)$$

where $\|\cdot\|$ denotes the usual Euclidean norm. An important property of $V(\alpha)$ is its invariance with respect to permutations of the data.

7.1. Interpolation and smoothing

It is interesting to recognize that the interpolation of a function, specified on a finite set of points, by means of a natural spline [19] is just an example of a C-generalized solution and that the smoothing of noisy data by means of splines [46] is just a regularized approximation of such a C-generalized solution.

The classical interpolation problem, namely the problem of determining a function f on $[0,1]$, given its values g_n on N points $0 \leq x_1 < x_2 < \dots < x_N \leq 1$

$$f(x_n) = g_n \ ; \ n = 1, \ldots, N \tag{7.25}$$

can be formulated in the form (7.1) if f belongs to a <u>reproducing kernel Hilbert space</u>, namely a Hilbert space of continuous functions such that all evaluation functionals are continuous. Then they take the form

$$f(x) = (f, Q_x)_X \tag{7.26}$$

and the function

$$Q(x,y) = (Q_x, Q_y)_X = Q_x(y) = Q_y(x) \tag{7.27}$$

is the <u>reproducing kernel</u> of X.

Assume that X is a space of functions with square integrable derivatives up to the order $k \geq 1$ and introduce in X the scalar product

$$(f, \phi)_X = \sum_{j=0}^{k-1} f^{(j)}(0)\, \phi^{(j)}(0) + \int_0^1 f^{(k)}(x) \phi^{(k)}(x)\, dx. \tag{7.28}$$

X is a reproducing kernel Hilbert space and, using Taylor formula for f(x), one gets that the reproducing kernel is

$$Q(x,y) = \sum_{j=0}^{k-1} (j!)^{-2} (x\, y)^j + \tag{7.29}$$

$$+ \sum_{j=0}^{k-1} (-1)^{k+j+1} \left[(2k-j-1)!\, j!\right]^{-1} x^{2k-j-1} y^j +$$

$$+ (-1)^k \left[(2k-1)!\right]^{-1} (x-y)_+^{2k-1}$$

where x_+ stands for the function which is equal to x when $x > 0$ and equal to zero when $x \leq 0$. Therefore the problem (7.25) can be formulated in X as a problem of the type (7.1) with $\phi_n(x) = Q(x_n, x)$. The generalized solution f^+ is just a linear combination of the ϕ_n and therefore it is a spline function. This spline is the function which minimizes the norm of f, as given by equation (7.28). Interpolation by means of a natural spline [19] results from the minimization of the seminorm

$$\| Cf \|_Z = \left(\int_0^1 |f^{(k)}(x)|^2 \, dx \right)^{1/2} \tag{7.30}$$

where $Z = L^2(0,1)$. The operator $C : X \longrightarrow Z$ is bounded, $\| C \| = 1$, and $N(C)$ is the linear subspace of the polynomials of degree k-1. Therefore, if $k \leqslant N$, the conditions i) - iii) of Sect. 3.4 are satisfied and there exists a unique C-generalized solution of the interpolation problem (7.25). Thanks to the well-known classical results of interpolation theory, this is the unique natural spline which interpolates the data.

On the other hand, the smoothing problem, as formulated by Reinsch [46], consists of minimizing the seminorm (7.30) with the constraint

$$\sum_{n=1}^{N} |f(x_n) - g_n|^2 \leqslant \varepsilon^2 \quad . \tag{7.31}$$

This is just a regularization method for the C-generalized solution previously discussed, with a choice of the regularization parameter provided by the discrepancy principle. As proved in [46] this regularized solution is still a spline function. Applications of the cross-validation method for the choice of the regularization parameter in this problem are given in [14].

7.2. Moment discretization of first kind Fredholm equations

Consider a first kind Fredholm integral equations with a continuous kernel $K(x,y)$

$$\int_a^b K(x,y) f(y) dy = g(x) \;\; ; \; c \leqslant x \leqslant d \quad . \tag{7.32}$$

If the intervals $[a,b]$ and $[c,d]$ are bounded, then the integral operator A associated with the continuous kernel $K(x,y)$ is a compact operator from $L^2(a,b)$ into $L^2(c,d)$. Denote its singular system by $\{ \lambda_k^{1/2}, u_k, v_k \}_{k=0}^{\infty}$. The method of moment discretization [40, 21] for the approximate solution of equation (7.32) is as follows: assume that g is known on a finite set of points, $c \leq x_1 < x_2 < \ldots < x_N \leq d$ and consider the problem of determining a function f satisfying the equations

$$\int_a^b K(x_n,y)\, f(y)\, dy = g_n; \; n = 1, \ldots, N \tag{7.33}$$

where $g_n = g(x_n)$. If $X = L^2(a,b)$, this is a problem of the type (7.1) with $\phi_n(y) = K(x_n,y)$.

We intend to investigate the relationship between the singular system of the integral operator A and the singular system of the operator A_N defined by equation (7.33). We will do it in the case where $x_1, x_2 .., x_N$ are equidistant points. If $w_1, w_2 .., w_N$ are the weights of some quadrature formula, we introduce in Y_N a scalar product defined by equation (7.13) with $W_{nn} = w_n$, $W_{nm} = 0$ when $n \neq m$. We also recall that the singular values of A are the square roots of the eigenvalues λ_k of the integral operator A*A whose kernel is given by

$$T(y,y') = \int_c^d K(x,y)\, K(x,y')\, dx \tag{7.34}$$

and that the singular functions u_k are the corresponding eigenfunctions. Analogously the singular values of A_N are the square roots of eigenvalues $\lambda_{N,k}$ of the finite rank integral operator $A_N^* A_N$ whose kernel is

$$T_N(y,y') = \sum_{n=1}^{N} w_n\, K(x_n,y)\, K(x_n,y') \tag{7.35}$$

(as follows from (7.13)) and the singular functions $u_{N,k}$ are the corresponding eigenfunctions. Clearly $T_N(y,y')$ is just the approximation of $T(y,y')$ provided by the quadrature formula corresponding to the abscissae x_n and weights w_n. From this remark one can derive the following result.

Theorem 7.1 - When $N \rightarrow +\infty$, the singular values $\lambda_{N,k}^{1/2}$ converge to the corresponding singular values $\lambda_k^{1/2}$ and the singular functions $u_{N,k}$ converge, in the norm of $L^2(a,b)$, to the corresponding singular functions u_k.

Proof: Consider the kernel $\tilde{R}_N(y,y') = T(y,y') - T_N(y,y')$. Since $K(x,y)$ is continuous, $\tilde{R}_N(y,y')$ tends to zero for any y,y' when $N \to \infty$. Furthermore $\tilde{R}_N(y,y')$ is continuous and bounded by a constant independent of N, since $K(x,y)$ is bounded and the sum of the weights is just the length of the interval $[c,d]$. Therefore, from the dominated convergence theorem it follows that

$$\lim_{N \to \infty} \int_a^b \left(\int_a^b |\tilde{R}_N(x,x')|^2 dx' \right) dx = 0 \quad . \tag{7.36}$$

The Weyl-Courant lemma on the perturbation of compact operators implies the inequality

$$|\lambda_k - \lambda_{N,k}| \leq \|\tilde{R}_N\| \tag{7.37}$$

where $\|\tilde{R}_N\|$ denotes the norm of the integral operator associated with the kernel $\tilde{R}_N(y,y')$. From equation (7.36) we get $\|\tilde{R}_N\| \longrightarrow 0$ and therefore the first part of the theorem is proved. The second part can be proved by investigating in a similar way the eigenprojections onto the subspaces spanned by the eigenfunctions [7]. □

The obvious consequence of this result is that the condition number of the problem (7.33) tends to infinity when $N \longrightarrow \infty$. Furthermore it can be proved (see, for instance [20]) that if $g \in R(A)$ and the g_n are the values of such a g at the points x_n, then the generalized solution (7.16) of the problem (7.33) converges in the norm of $L^2(a,b)$ to the generalized solution of the integral equation (7.32). This result can also be interpreted as a regularization, through discretization of the data, of the solution of first kind Fredholm integral equations.

REFERENCES

1. BACKUS, G. and GILBERT, F., 1968, The resolving power of gross earth data, Geophys. J.R. Astr. Soc., 16, 169-205
2. BALTES, H.P. (ed.), 1978, Inverse Source Problems in Optics, Topics in Current Physics, Vol 9 (Springer, Berlin)
3. BALTES, H.P. (ed.), 1980, Inverse Scattering Problems in Optics,

Topics in Current Physics, Vol 20 (Springer, Berlin)

4. BAKUSHINSKII, A.B., 1965, A numerical method for solving Fredholm integral equations of the first kind, USSR Comp. Math. Math.Phys. 5 (No.4), 226-233

5. BERTERO, M. and VIANO, G.A., 1978, On probabilistic methods for the solution of improperly posed problems, Bollettino U.M.I. 15-B, 483-508

6. BERTERO, M., 1982, Problemi lineari non ben posti e metodi di regolarizzazione. (Consiglio Nazionale delle Ricerche, Firenze)

7. BERTERO, M., DE MOL, C. and PIKE, E.R., 1985, Linear inverse problems with discrete data. I: General formulation and singular system analysis, Inverse Problems 1, 301-330

8. BOERNER, W.M., BRAND, H., CRAM, L.A., GJESSING, T.D., JORDAN, A.K., KEYDEL, W., SCHWIERZ, G. and VOGEL, M. (eds.), 1985, Inverse Methods in Electromagnetic Imaging, Part 1 and Part 2 (Reidel, Dordrecht)

9. CARASSO, A. and STONE, A.P., (eds.), 1975, Improperly Posed Boundary Value Problems (Pitman Publ., London)

10. CHADAN, K. and SABATIER, P.C., 1977, Inverse Problems in Quantum Scattering Theory (Springer, Berlin)

11. CIULLI, S., POMPONIU, C. and SABBA-STEFANESCU, I, 1975, Analytic extrapolation techniques and stability problems in dispersion relation theory , Phys. Reports, 17, 133-224

12. COLIN , L. (ed.), 1972, Mathematics of Profile Inversion, Proc. of a workshop held at Ames Research Center, Moffat Field, CA 994084, June 12-16, 1971; NASA TM-X-62. 150

13. COLLI FRANZONE, P., GUERRI, L., TACCARDI, B. and VIGANOTTI, C., 1985, Finite element approximation of regularized solutions of the inverse potential problem of electrocardiography and applications to experimental data, Calcolo, 22, 91-186

14. CRAVEN, P. and WAHBA G., 1979, Smoothing noisy data with spline functions: estimating the correct degree of smoothing by the method of generalized cross validation, Numer.Math., 31, 377-403

15. FOX, D. and PUCCI, C., 1958, The Dirichlet problem for the wave equation, Ann.Mat.Pura Appl. 46, 155-182

16. FRANKLIN, J.N., 1970, Well-posed stochastic extensions of ill-posed linear problems, J.Math. Anal.Appl. 31, 682-716

17. FRANKLIN, J.N., 1974, On Tikhonov's method for ill-posed problems, Math. Comp. 28, 889-907

18. GREGORY, R.T. and KARNEY, D.L., 1969, A Collection of Matrices for Testing Computational Algorithms (Wiley-Interscience, New York)

19. GREVILLE, T.N.E. (ed.), 1969, Theory and Applications of Spline Functions (Academic Press, New York)

20. GROETSCH, C.W., 1977, Generalized Inverses of Linear Operators

(Dekker, New York)

21. GROETSCH, C.W., 1984, The Theory of Tikhonov Regularization for Fredholm Equations of the First Kind, Research Notes in Mathematics, Vol.105 (Pitman, Boston)

22. HADAMARD, J., 1923, Lectures on the Cauchy Problem in Linear Partial Differential Equations (Yale University Press, New Haven)

23. HERMAN, G.T., 1980, Image Reconstruction from Projections: The Fundamentals of Computerized Tomography, (Academic Press, New York)

24. HILLE, E. and TAMARKIN, J.D., 1931, On the characteristic values of linear integral equations, Acta Math. 57, 1-76

25. HUANG, T.S., (ed.), 1975, Picture Processing and Digital Filtering, Topics in Applied Physics, Vol.6 (Springer, Berlin)

26. IVANOV, V.K., 1962, On linear problems which are not well-posed, Soviet Math.Dokl. 3, 981-983

27. IVANOV, V.K., 1966, The approximate solution of operator equations of the first kind, USSR Comp. Math. Math.Phys. 6, (No.6) 197-205

28. JOHN, F., 1955, Numerical solution of the equation of heat conduction for preceding times, Ann.Mat.Pura Appl. 40, 129-142

29. JOHN, F., 1960, Continuous dependence on data for solutions of partial differential equations with a prescribed bound, Comm. Pure Appl. Math. 13, 551-585

30. KAMMERER, W.J. and NASHED, M.Z., 1971, Steepest descent for singular linear operators with nonclosed range, Applicable Analysis, 1, 143-159

31. KAMMERER, W.J. and NASHED, M.Z., 1972, On the convergence of the conjugate gradient method for singular linear operator equations, SIAM J.Numer.Anal., 9, 165-181

32. LANDWEBER, L., 1951, An iteration formula for Fredholm integral equations of the first kind, Amer.J.Math. 73, 615-624

33. LAVRENTIEV, M.M., 1967, Some Improperly Posed Problems of Mathematical Physics (Springer, Berlin)

34. LOUIS, A.K. and NATTERER, F., 1983, Mathematical problems of computerized tomography,Proc. IEEE, 71, 379-389

35. MILLER, K., 1964, Three circle theorems in partial differential equations and applications to improperly posed problems, Arch.Rat.Mech.Anal. 16, 126-154

36. MILLER, K., 1970, Least squares methods for ill-posed problems with a prescribed bound, SIAM J. Math.Anal. 1, 52-74

37. MOROZOV, V.A., 1968, The error principle in the solution of operational equations by the regularization method, USSR Comp.Math.Math.Phys. 8 (No.2), 63-87

38. MOROZOV, V.A., 1984, Methods for Solving Incorrectly Posed Problems (Springer, Berlin)

39. NASHED, M.Z. (ed.), 1976, Generalized Inverses and Applications (Academic Press, New York)

40. NASHED, M.Z., 1976, On moment-discretization and least-squares solutions of linear integral equations of the first kind, J.Math.Anal.Appl., 53, 359-366

41. PAPOULIS, A., 1968, ,Systems and Transforms with Applications in Optics (McGraw-Hil, New York)

42. PAYNE, L.E., 1975, Improperly Posed Problems in Partial Differential Equations, Regional Conf.Series in Apllied Math. (SIAM)

43. PHILLIPS, D.L., 1962, A technique for the numerical solution of certain integral equations of the first kind, J.Asso.Comput.Mach. 9, 84-97

44. PICARD, E., 1910, Sur un théorème géneral relatif aux equations integrales de première espèce et sur quelques problèmes de physique mathèmatique, R.C.Mat.Palermo, 29, 615-619

45. PUCCI, C., 1955, Sui problemi di Cauchy non "ben posti", Atti Acc.Naz.Lincei, 18, 473-477

46. REINSCH, C.H., 1967, Smoothing by spline functions, Numer.Math., 10, 177-183

47. ROBINSON, E.A., 1982, Spectral approach to geophysical inversion by Lorentz, Fourier and Radon transforms, Proc.IEEE, 70, 1039-1054

48. SABATIER, P.C. (ed.), 1978, Applied Inverse Problems, Lecture Notes in Physics, Vol.85 (Springer, Berlin)

49. SLEPIAN, D. and POLLAK, H.O., 1961, Prolate spheroidal wave functions, Fourier analysis and uncertainty - I, Bell System Tech. J. 40, 43-64

50. TALENTI, G., 1978, Sui problemi mal posti, Bollettino U.M.I. 15-A, 1-29

51. TIKHONOV, A.N., 1963, Solution of incorrectly formulated problems and the regularization method, Soviet Math.Dokl. 4, 1035-1038

52. TIKHONOV, A.N., 1963, Regularization of incorrectly posed problems, Soviet Math.Dokl. 4, 1624-1627

53. TIKHONOV, A.N. and ARSENIN, V.Y., 1977, Solutions of Ill-Posed Problems, translation ed. by F.John (Winston/Wiley, Washington)

54. TITCHMARSH, E.C., 1948, Introduction to the Theory of Fourier Integrals (Oxford University Press)

55. WAHBA, G., 1977, Practical approximate solutions to linear operator equations when the data are noisy, SIAM J.Numer.Anal., 14, 651-667

56. WESTWATER, E.R., SNIDER, J.B. and CARLSON, A.V., 1975, Experimental determination of temperature profiles by ground based microwave radiometry, J.Appl.Meteor., 14, 524-539

SOME MATHEMATICAL PROBLEMS MOTIVATED BY MEDICAL IMAGING

F. Alberto Grünbaum

Department of Mathematics
University of California
Berkeley, California 94720

LECTURE 1: Limited Data Reconstruction Problems

Consider the problem of recovering a function $f(\underline{x})$, $\underline{x} \in R^n$ from certain of its one dimensional projections

$$Pf(\underline{\omega},t) = \int_{\langle\underline{x},\underline{\omega}\rangle=t} f = \int_{R^n} f(\underline{x})\delta(t - \langle\underline{x},\underline{\omega}\rangle) , \qquad \underline{\omega} \in S^{n-1} , \quad t \in R .$$

In the expression above $\langle\underline{x},\underline{\omega}\rangle$ denotes the inner product of a vector $\underline{x} \in R^n$ and a unit vector $\underline{\omega}$ in R^n, with S^{n-1} denoting the unit sphere in R^n.

This problem arises in a variety of fields ranging from wave propagation, statistics, geophysics, X-ray medical imaging and magnetic resonance imaging. More of these applications probably still lie in the future as people working in different fields find ways of ***interpreting*** certain measurements pertaining to an unknown function f as related to such "sections" or "tomos" through the "object". For the first work in this direction credit belongs to H. Lorentz in the 3-dimensional case (see [1]), and to J. Radon in the general case of arbitrary dimension [2]. For a general review of the field the reader can consult [2,3,4,5,6].

To set up the scene for our considerations, observe the so called "central projection theorem" relating the one-dimensional Fourier transform of $Pf(\omega,\cdot)$ to the n-dimensional Fourier transform of $f(x)$. If the latter is defined by

$$\hat{f}(\underline{y}) = \int_{R^n} f(\underline{x})\, e^{-i\langle\underline{y},\underline{x}\rangle 2\pi}\, d\underline{x}$$

and we put $\underline{y} = |y|\underline{\omega}$, $|\underline{\omega}| = 1$, we can reexpress the integral above as

$$\begin{aligned}\hat{f}(y) &= \int_{R^n} e^{-2\pi i|y|\langle\underline{\omega},\underline{x}\rangle} f(\underline{x})d\underline{x} \\ &= \int_{-\infty}^{\infty} e^{-2\pi i|y|t}\, dt \int_{\langle\underline{x},\underline{\omega}\rangle=t} f(\underline{x}) \\ &= \int_{-\infty}^{\infty} e^{-2\pi i|y|t}\, Pf(\underline{\omega},t)\, dt \quad .\end{aligned}$$

The meaning of this is clear: the knowledge of $Pf(\underline{\omega},\cdot)$ is equivalent to that of $\hat{f}(y)$ along the line through the origin given by

$$\underline{y} = \underline{\omega}|y| \quad .$$

It is clear that these lines fill up Fourier space and thus assuming all the technical conditions amply described in the literature one sees that the knowledge of $Pf(\underline{\omega},t)$ for all $\underline{\omega} \in S^{n-1}$, $t \in R$, determines the function $\hat{f}(\underline{y})$ and thus $f(\underline{x})$.

The description above is not rich enough to cover any of the cases encountered in "practice". A more realistic problem consists of assuming that we have "noisy" measurements of a finite number of the projections

$$Pf(\underline{\omega}_k, t_j) \quad , \qquad \begin{matrix} k = 1,\ldots,N \;, \\ j = 1,\ldots,M \;. \end{matrix}$$

Each one of these two discretizations brings problems of its own. A careful treatment of the situation when only the $\underline{\omega}_k$ discretization is dealt with, can be found in [5]. The bottom line of the analysis in [5] is that one can give sharp error bounds relating constructs built out of the data and a "mollified" or "smoothed out" version of $f(\underline{x})$. Any attempt to get $f(\underline{x})$ itself, i.e., pointwise values, is seen to be doomed. This answers a paradox pointed out in [6]. It is appropriate to mention that a similar analysis for the "fan beam geometry" dominant among clinical machines in the market for the last few years has not been carried out, mainly because of mathematical difficulties.

It is noteworthy to mention that the clinical field has driven the development of the equipment - and the difficulty of the corresponding mathematical problems - at a fascinating rate: in 1972 it took G.Hounsfield (EMI) about 5 minutes for one 2-dimensional cross section; in 1986 the machine developed by D.Boyd and others (IMATRON) takes .04 seconds for the same job. This latter machine is an example of what is called "limited angle tomography": one only has available $Pf(\underline{\omega},t)$ for $\underline{\omega}$ restricted to a certain sector in the unit circle. If one keeps in mind the "central section theorem" mentioned earlier we see that we are dealing with the following general problem.

Given an unknown function $f(\underline{x})$, $\underline{x} \in R^n$, determine f from the knowledge of $\hat{f}(\underline{y})$ for $\underline{y}$ in a certain subset A of R^n.

It is clear that this problem does not have a unique solution unless some *a priori* knowledge about f(x) is supplied. In our case we know (in most applications at least) that $f(\underline{x})$ has a bounded, and known, support; denoted by B.

A better formulation of the problem is therefore embodied in the pair of equations

$$Bf = f \qquad \text{a priori information} \quad ,$$

$$AFf = g \qquad \text{data} \quad .$$

In the equations above, A and B denote by abuse of notation the operators of restriction of a function to the sets A and B respectively. F denotes the Fourier transform.

If one combines these two equations we are faced with the equation for f given by

$$AFBf = g \quad .$$

There is a standard method for analyzing such a problem: one looks for the singular value decomposition of the linear operator AFB and thus obtains a handle on the inversion problem. All questions about ill conditioning, stability, etc., are revealed by a ***close*** look at the singular values and singular vectors of AFB. The latter turn out to be the eigenvectors of $(AFB)(AFB)^*$ and $(AFB)^*(AFB)$, and their eigenvalues give the (squares of the) singular values.

An example of the usefulness of these concepts is given by the following analysis of the Gerchberg-Saxton-Papoulis algorithm used in electron microscopy [7,9] as well as in tomography [8] and myriad other applications.

If u_i denotes the singular functions of AFB and μ_i are the singular values, one can easily see that for

$$f = \sum_{i=1}^{\infty} c_i u_i$$

one has

$$f^{(n)} = \sum_{i=1}^{\infty} (1 - (1 - \mu_i^2)^n) c_i u_i$$

where $f^{(n)}$ denotes the n-th iterate of the algorithm obtained by going back and forth between physical and Fourier space using every time all the information available.

Since one can prove mathematically that for (bounded) A,B the μ_i satisfy

$$0 < \mu_i < 1$$

it is clear that one should have

$$f^{(n)} \longrightarrow f \quad \text{as} \quad n \quad \text{goes to infinity.}$$

In practice, however, one observes that

$$f^{(n)} \longrightarrow \sum_{i=1}^{N} c_i u_i$$

where N is independent of f, but depends on A and B.

In the case when A and B are intervals in the real line one can go ***much further*** and see that only the first N among the μ_i's are substantially different from zero, thus explaining the phenomenon mentioned earlier.

Such a detailed analysis is made possible by a remarkable series of ***miracles*** uncovered and exploited in the context of communication engineering by D.Slepian, H.Landau and H.Pollak in the early sixties.

This will be explained in Lecture #2.

A final point: there is little question that mathematics has made an important contribution to tomography and other areas of inverse problems. These lectures will try to travel the reverse path showing a number of interesting mathematical problems that one probably would not have encountered if it were not for some applied area posing the question.

LECTURE 2: Time and Band Limiting

Let A be the interval $[-\Omega,\Omega]$ in "frequency space" R^1 and B the interval $[-T,T]$ in physical space R^1. Let F be the Fourier transform from $L^2(R^1)$ to $L^2(R^1)$. Then the operator

$$(AFB)^*(AFB)$$

mentioned last time is easily seen to be given by a finite convolution, namely

$$(Kh)(x) = \int_{-T}^{T} \frac{\sin \Omega(x-y)}{x-y} h(y)dy \quad , \qquad -T \leq x \leq T .$$

To see this, recall that AFB itself is given by

$$((AFB)h)(\lambda) = \int_{-T}^{T} e^{i\lambda y} h(y)dy \quad , \qquad -\Omega \leq \lambda \leq \Omega$$

and observe that the kernel in the convolution is given by

$$k(x-y) = \int_{-\Omega}^{\Omega} e^{i\lambda x} e^{-i\lambda y} d\lambda \quad ,$$

i.e., k is built from the eigenfunctions $\phi(x,\lambda) = e^{i\lambda x}$ of $-d^2/dx^2$ by means of

$$k(x-y) = \int_{-\Omega}^{\Omega} \phi(x,\lambda) \overline{\phi(y,\lambda)} d\lambda \ .$$

As remarked last time it is very useful to obtain an accurate and economical way of computing lots of eigenfunctions of the finite convolution operator given above. You may ask: how many such eigenfunctions do you really want, and why bother with a sophisticated method?

I can think of two answers: a) Quite often the point of this computation is not a reconstruction based on these eigenfunctions but insight into the dependence of "quality of information" in terms of "amount of data". In these cases you would like all of them, along with the eigenvalues. b) If one is going to use these functions in an actual reconstruction in medical tomography, recall that any decent screen has 512 x 512 pixels: that is a lot of degrees of freedom.

If we consider that b is the data vector, typically of dimension 180 N, with N the number of detectors, and x is the unknown density f and that these quantities are related by

$$Ax = b \quad .$$

we know that x is given by

$$x = \sum_{i=1}^{M} \frac{1}{\lambda_i} (b,v_i)u_i$$

with u_i, v_i, λ_i the singular system of A.

For an account of this one can see the accompanying lectures by M.Bertero and F.Natterer.

Since the reconstructions in medical imaging aim at a very high resolution, the value of M above should be rather large, and if one were to use this method for a practical reconstruction – a rather poor proposal – one could not rely on the power method or the like to compute directly the eigenfunctions of A^*A.

So, we charge on, looking for a method to compute these eigenfunctions.

For the problem at hand – a caricature to be sure – we do not even have to be clever: the work of Slepian, Landau and Pollak [10-14] is ready made for us. They found out that the differential operator

$$Dh \equiv ((T^2 - x^2)h')' - \Omega^2 x^2 h \quad , \qquad -T \leq x \leq T$$

with boundary conditions given by the requirement that its eigenfunctions should remain bounded at both end points, is a selfadjoint operator which commutes with the K given earlier. Since D has simple spectrum, each eigenfunction of D is automatically one of K. Indeed, if

$$\text{then} \quad \begin{aligned} Dv &= \mu v \\ KDv &= DKv = \mu Kv \end{aligned}$$

and since μ is a simple eigenvalue, $Kv = \lambda v$.

If you are new to this circle of ideas you may not realize that you have witnessed a miracle: you can count with few fingers the cases where this remarkable commutativity holds between an operator of the form

$$(AFB)^*(AFB) \ , \qquad A \subset X \ , \quad B \subset Y$$

and a differential operator on the space of functions on A. Here A and B are subsets of two spaces X,Y and functions on them are related by F.

The fact that D is "tridiagonal" is only half of the story: as it turns out the problem of computing the eigenvectors of D is numerically ***well*** conditioned while that of computing those of K is terribly ***ill posed***. This is true on top of the question of storage and computing time involved in doing the reduction to tridiagonal form that would be required in handling K.

Is this an important point or just a playground for mathematicians? I will let D.Slepian (not a professional mathematician by his own description) answer the question. I quote from his J. von Neumann lecture, (see [15]).

> I am going to tell you in detail about a problem in Fourier analysis that arose in a quite natural manner in a corner of electrical engineering known as Communication Theory. The problem was first attacked more than 20 years ago, jointly by myself and two colleagues at Bell Labs, Henry Pollak and Henry Landau. It differed from other problems I have worked on in two fundamental ways. First we solved it - completely, easily and quickly. Second, the answer was interesting - even elegant and beautiful. (Usually I struggle for months or years with a problem. If I do "solve" it, it is usually only in part and the answer itself is rarely interesting. The interest generally lies in the fact that I have proved that this is the answer.) In the case of this problem, however, the answer had so much unexpected structure that we soon saw that we had solved many other problems as well. We had answered questions we

had not meant to ask in optics, estimation and detection theory, quantum mechanics, laser modes — to name a few.

There was a lot of serendipity here, clearly. And then our solution, too, seemed to hinge on a lucky accident - namely that we found a second-order differential operator that commuted with an integral operator that was at the heart of the problem. Soon afterwards, a number of obvious generalizations of the original problem yielded to the same techniques. They had answers with the same elegant structure, and again there was by good luck a commuting second-order differential operator.

We had scratched the ground a bit and had unknowingly uncovered the tip of a rich vein of ore. Off and on for the next 20 years I came back and mined a new piece of it. Nor is it exhausted yet. In recent years Professor F. A. Grünbaum and his students have dug their shovels in and unearthed interesting novel generalizations and ramifications of the original problem. The mystery of this serendipity grows. Most of us feel that there is something deeper here than we currently understand - that there is a way of viewing these problems more abstractly that will explain their elegant solution in a more natural and profound way, so that these nice results will not appear so much as a lucky accident.

Another example considered by Slepian starts from

$$L = -\frac{d^2}{dx^2} + \frac{\nu(\nu+1)}{x^2} \quad \text{with} \quad \phi(x,k) = J_\nu(xk)\sqrt{xk}$$

and comes about by replacing R^1 by higher dimensional Euclidean space. For a fuller account, (see [13]).

Here is another example, or family of examples actually. Consider the operator

$$L \equiv -\frac{d^2}{dx^2} + x^2$$

in place of $-d^2/dx^2$. Decompose $L^2(R)$ in terms of the eigenfunctions of L, the Hermite functions $h_n(x)$. Notice that "frequency space" has now become discrete, $n=0,1,2,\ldots$. The analog of $(AFB)^*(AFB)$ is the $N\times N$ matrix

$$G_{ij}(T) = \int_{-\infty}^{T} h_i(\xi)h_j(\xi)d\xi$$

or the integral operator in $(-\infty,T)$ with kernel given by $K_N(x,y) = \sum_{k=0}^{N} h_k(x)h_k(y)$. We have, (see [16]),

THEOREM. *For every* N,T *there exists a tridiagonal matrix with simple spectrum which commutes with* $G(T)$, *and a second order differential operator commuting with* K_N.

How far can this be pushed? A good open problem. So far it is known to hold when one replaces the Hermite operator by any of those corresponding to the classical orthogonal polynomials: Jacobi, Laguerre, Bessel, Hahn, Krawtchouk, Askey-Wilson, ...

See [16,17,18,19]. It fails for any example that I have tried outside of the list given above.

One may observe that in all the examples just mentioned, the eigenfunctions of $L = -\frac{d^2}{dx^2} + V(x)$ satisfy either a second order differential equation in the spectral parameter k, or a 3-term recursion relation in n, i.e., either

$$L\phi = k^2\phi$$

with

$$\left(\sum_{i=0}^{2} a_i(k)\partial_k^i\right)\phi(x,k) = \Xi(x)\phi(x,k)$$

or

$$L\phi_n = c_n\phi_n$$

with

$$\alpha_n\phi_{n-1}(x) + \beta_n\phi_n(x) + \gamma_n\phi_{n-1}(x) = \Xi(x)\phi_n(x) \quad .$$

It appears that the existence of this "second equation" plays a crucial role in obtaining a commuting second order differential operator. This warrants a more detailed analysis of the existence of such a "second equation", as is done in [20]. We return to this point at the end of the next lecture.

For a final example, of a different nature, think of the N×N Hilbert matrix

$$H^N_{ij} = \frac{1}{i + j - 1} \quad , \qquad i,j = 1,\ldots,N \quad .$$

One can prove (see [21]) that the tridiagonal matrix given

$$(T^N)_{ii} = -2(N-i)(N+i-1)(i^2-i-N+1)$$
$$(T^N)_{i,i+1} = (T^N)_{i+1,i} = i^2(N^2-i^2)$$

satisfies

$$T^N H^N = H^N T^N \quad .$$

Here we see the same features exhibited by the examples above: not only is T^N tridiagonal, but it has ***all*** its eigenvalues very separated. For $N \approx 10$ the advantage in replacing H^N by T^N – when computing its eigenvectors – is only apparent in reference to the eigenvalues of H^N ***clustered*** around zero. For $N \approx 10^4$ the advantage extends to all the eigenvectors because of storage and computing time.

As it turns out, this example led to the extension of this circle of ideas for the Laplace transform, as will be seen next time.

LECTURE 3: Commuting Integral and Differential Operators

The essential ingredients so far have been three operators A,B,F: restrictions in frequency and physical space and the Fourier transform.

One can generalize the situation by considering F to be a map

$$(Ff)(k) = \int \overline{\Phi(x,k)}\, f(x)dx$$

where $\Phi(x,k)$ are the eigenfunctions of a second order differential operator of the form

$$L = -\frac{d^2}{dx^2} + V(x)$$

so that we've

$$L\Phi(x,k) = k^2\Phi(x,k) .$$

For $V(x) = 0$ we get $\Phi(x,k) = e^{ixk}$ and the usual Fourier transform. For $V(x) = x^2$ we get the Hermite example considered last time. For $V(x) = \frac{c_n}{x^2}$ (with an appropriate choice of c_n) we get the examples obtained by Slepian [19] for R^n, $n \geq 1$.

In fact we do not need L to be a second order differential operator, although all the examples so far can be put under this umbrella.

The operator AFB acts according to

$$(AFB\ f)(k) = \begin{cases} \int_B \overline{\Phi(y,k)}f(y)dy \\ 0 \quad \text{if} \quad k \notin A \end{cases}$$

and thus we get that since

$$(AFB)^*(AFB) = BF^*AFB \qquad (A^2 = A)$$

and

$$(F^*g)(x) = \int \Phi(x,k)g(k)dk$$

we finally obtain

$$((AFB)^*(AFB)f)(x) = \begin{cases} \int_A \Phi(x,k)dk \int_B \overline{\Phi(y,k)}f(y)dy \\ \text{for} \quad x \in B \\ 0 \quad \text{otherwise} \end{cases}$$

$$= \int (\int \Phi(x,k)\overline{\Phi(y,k)}dk)f(y)dy$$

$$\equiv \int_B K_A(x,y)f(y)dy .$$

Here

$$K_A(x,y) \equiv \int_A \Phi(x,k)\overline{\Phi(y,k)}dk .$$

To continue with our theme we consider, for an arbitrary "resolution of the identity"

$$\Phi(x,k) , \qquad -\infty < k < \infty$$

such that

$$\int_{-\infty}^{\infty} \Phi(x,k)\overline{\Phi(y,k)}dk = \delta(x-y)$$

the problem of finding, for any choice of sets A and B, a differential operator commuting with the integral operator that acts on the set B via the integral kernel given by $K_A(x,y)$.

Maybe one more example is useful at this point. Consider the Laplace transform

$$(\pounds f)(p) = \int_0^{\infty} e^{-pt} f(t)dt$$

acting from $L^2(0,\infty)$ to $L^2(0,\infty)$. We take now $B \equiv (a,b) \subset (0,\infty)$ and $A \equiv (p_1,p_2) \subset (0,\infty)$ and we get

$$((A\pounds B)f)(p) = \int_a^b e^{-pt} f(t)dt , \qquad p_1 \leq p \leq p_2 , \quad 0 \text{ otherwise}$$

and

$$((A\pounds B)^*g)(t) = \int_{p_1}^{p_2} e^{-tp} g(p)dp , \qquad a \leq t \leq b , \quad 0 \text{ otherwise} .$$

Just as before we get

$$((A\pounds B)^*(A\pounds B))f(t) = \int_a^b \Big(\int_{p_1}^{p_2} e^{-pt} e^{-ps} dp\Big)f(s)ds$$

$$= \int_a^b \frac{e^{-p_1(t+s)} - e^{-p_2(t+s)}}{t + s} f(s)ds .$$

Notice also that, by a trivial exchange

$$((A\pounds B)(A\pounds B)^*)g(p) = \int_{p_1}^{p_2} \frac{e^{-a(p+q)} - e^{-b(p+q)}}{p + q} g(q)dq .$$

In the special case $p_1 = 0$, $p_2 = \infty$, the first operator becomes

$$(Kf)(t) = \int_a^b \frac{f(s)}{t+s} ds$$

and the second one gives

$$(Mg)(p) = \int_0^{\infty} \frac{e^{-a(p+q)} - e^{-b(p+q)}}{p + q} g(q)dq .$$

In this case it is also possible to find a differential operator commuting with K and one commuting with M. They are given by

$$D = \frac{d}{dt}(t^2 - a^2)(b^2 - t^2)\frac{d}{dt} - 2(t^2 - a^2)$$

and

$$\tilde{D} = -\frac{d}{dp^2}p^2\frac{d}{dp^2} + (a^2 + b^2)\frac{d}{dp}p^2\frac{d}{dp} - a^2b^2p^2 + 2a^2$$

respectively. For a complete discussion, see [22].

Suppose we are given an integral operator of the form

$$(Kf)(t) = \int_a^b k(t,s)f(s)ds$$

and that we set out to find a differential operator of the form

$$D = \frac{d}{dt}\alpha(t)\frac{d}{dt} + \beta(t) \quad , \qquad a \leqslant t \leqslant b$$

such that

$$KD = DK \quad .$$

We leave for later the question of the appropriate domain in which the symmetric differential expression D is to be considered as a self-adjoint operator.

The following computation is crucial.

$$((DK)f)(t) = \int_a^b \frac{d}{dt}\alpha(t)\frac{d}{dt}k(t,s)f(s)ds + \int_a^b \beta(t)k(t,s)f(s)ds \quad .$$

On the other hand, $(KDf)(t)$ is given by

$$\int^b k(t,s)\left(\frac{d}{ds}\alpha(s)\frac{d}{ds}f(s) + \beta(s)f(s)\right)ds$$

$$= k(t,s)\alpha(s)\frac{df}{ds}\Big|_{s=a}^{s=b} - \left|\alpha(s)\frac{d}{ds}k(t,s)f(s)\right|_{s=a}^{s=b}$$

$$+ \int_a^b \left(\frac{d}{ds}\alpha(s)\frac{d}{ds}k(t,s) + \beta(s)k(t,s)\right)f(s)ds$$

and if this is to agree with $(DKf)(t)$ for a dense set of functions in $L^2(a,b)$ we need a couple of things:

(a) $D_t k(t,s) = D_s k(t,s)$

to insure that the integral terms should balance, and

(b) the vanishing of the boundary terms.

This second point warrants a bit more discussion. We could restrict the domain of D by asking that $f(a) = f(b) = f'(a) = f'(b) = 0$ but this would be too many conditions for a second order operator. We are naturally led to requiring that α should vanish at the end points $\alpha(a) = \alpha(b) = 0$. If these zeros are simple and β is regular at a,b one easily sees that the indicial equations for $(D - \lambda I)$ has $\rho=0$ as a double root for any λ. Thus for each λ there is only one bounded solution around each singular point and one can take as domain of D the set of functions f in $L^2(a,b)$ which are absolutely continuous, have Df in $L^2(a,b)$ and tend to finite limits as t approaches either a or b. A prototype of this situation is the Legendre differential equation with these standard boundary conditions. For a fuller discussion of this circle of ideas, see [23].

Notice that in the case of the Fourier and the Laplace transform mentioned earlier we get

$$\alpha(t) = T^2 - t^2 \qquad -T \leqslant t \leqslant T$$

$$\alpha(t) = (t^2 - a^2)(b^2 - t^2) \qquad a \leqslant t \leqslant b$$

with $\beta(t)$ regular at the end points of the interval.

Recall that in the Laplace case the relevant functions $\Phi(x,k)$ are

$$e^{-pt}$$

satisfying the pair of equations

$$Le^{-pt} \equiv \frac{d^2}{dt^2} e^{-pt} = p^2 e^{-pt} ,$$

$$Be^{-pt} \equiv \frac{d^2}{dp^2} e^{-pt} = t^2 e^{-pt} = \Xi(t)e^{-pt} .$$

If we look for the commuting differential operator D in the form

$$c_1 \Xi L \Xi + c_2(L\Xi + \Xi L) + c_3 L + \text{lower order}$$

we see from

$$L = d^2/dt^2 , \qquad \Xi = t^2 ,$$

$$\Xi L \Xi = t^4 \frac{d^2}{dt^2} + 4t^3 \frac{d}{dt} + 2t^2$$

$$L\Xi + \Xi L = 2t^2 \frac{d^2}{dt^2} + 4t \frac{d}{dt} + 2$$

that the requirement

$$D = (t^2 - a^2)(b^2 - t^2) \frac{d^2}{dt^2} + \text{lower order}$$

forces the choices

$$c_1 = -1 , \qquad c_2 = \frac{a^2 + b^2}{2} , \qquad c_3 = -a^2b^2 .$$

It is now easy to see that the resulting

$$c_1 \Xi L \Xi + c_2(L\Xi + \Xi L) + c_3 L$$

agrees with the operator D given earlier up to addition of the term $(a^2-b^2)I$. The idea of expressing D in terms of these operators made up of L and Ξ is suggested by ideas in R.Perline's [19] recent work.

A final point is the question: when can one have a Schroedinger, or higher order, differential operator L such that a family of its eigenfunctions $\phi(x,k)$ satisfies the "equation in the spectral parameter"

$$B\phi \equiv \left(\sum_{i=0}^{m} a_i(k) \frac{\partial^i}{\partial k^i}\right)\phi(x,k) = \Xi(x)\phi(x,k) .$$

This is exhaustively dealt with in [20]. Here we just mention that for $L = -\frac{d^2}{dx^2} + V(x)$ it turns out that one half of the cases when both equations hold are given by choosing $V(x)$ such that

(a) $V(x)$ is rational,

(b) $V(x,t)$ is rational for all $t \geq 0$ and

$V(x,t)$ is the solution of the Korteweg-de Vries equation

$$V_t = V_{xxx} - 6VV_x$$

with initial data $V(x)$.

We remark that the corresponding operators $L \equiv -\frac{d^2}{dx^2} + V(x)$ can be obtained by starting from $L_0 = -\frac{d^2}{dx^2}$ and performing a finite number of Darboux factorizations into first order differential operators

$$\begin{aligned} L_0 &= A_0^* A_0 \\ L_1 &\equiv A_0 A_0^* = A_1^* A_1 \\ &\cdots\cdots \\ &\cdots\cdots \\ L_n &\equiv A_{n-1} A_{n-1}^* \equiv L \end{aligned} .$$

The reader may see some similarity with the QR factorization process used to approximate eigenvectors, which is covered in F.Natterer's lectures in this volume.

The method described above produces one half of the solutions to our problem: what about the other half? Answer: Apply once again the Darboux process but start from a different L_0, namely $-\frac{d^2}{dx^2} - \frac{1}{4x^2}$.

More on this Darboux (also called factorization process in the physics literature) method at the end of the next lecture. This method should be better known and we give a short account of it.

Start from L and put

$$L = -D^2 + V = (-D + U)(D + U) = -D^2 - U' + U^2 .$$

This means that

$$V = U^2 - U'$$

i.e., a Riccati differential equation. Put $U = +\phi'/\phi$ to get

$$V = \frac{\phi'^2}{\phi^2} + \frac{\phi''}{\phi} - \frac{\phi'^2}{\phi^2} = \frac{\phi''}{\phi}$$

i.e., ϕ should be ***any*** eigenfunction of L with $L\phi = 0$.

Since these ϕ form a two-dimensional space and the operation ϕ'/ϕ throws out one degree of freedom, we see that the ***new*** operator

$$L_1 \equiv (D + U)(-D + U) = -D^2 + \tilde{V}$$

is really a (complex) one-dimensional family of operators

$$L_1(t) = -D^2 + \tilde{V}(x,t) , \qquad t \in C .$$

LECTURE 4: Magnetic Resonance Imaging

Over the last few years there has been great progress in the area of medical imaging based on magnetic resonance. This is based in the use of large superconducting magnets with a gradient in the field direction which enable encoding of spatial information. By the application of a radio frequency field $(B_1(t), B_2(t))$, $0 \leq t \leq T$, in the plane orthogonal to the gradient and a discrimination of the amount of spins precesing at a given frequency, one can measure "plane integrals" of the unknown spin density and then perform a reconstruction using tomographic ideas. Naturally this is neither the only nor the best way to perform the reconstruction.

An important problem is that of designing applied fields $(B_1(t), B_2(t))$, $0 \leq t \leq T$ such that the magnetization vector $\bar{M}(T, x,y,z)$ (T is the final time before readings start) has some desirable properties.

The dynamics are governed by the Bloch equations

$$\dot{\bar{M}} = \gamma\bar{M} \times \bar{B}$$

$$\bar{B}(t,x,y,z) \equiv \bar{B}(t) = (B_0 + Gz)\bar{k} + B_1(t)\bar{i} + B_2(t)\bar{j}$$

with initial condition, as an example not to be changed for this lecture,

$$\bar{M}(0,x,y,z) = \begin{pmatrix} 0 \\ 0 \\ -1 \end{pmatrix} \equiv \bar{M}(0) .$$

Notice that since $\bar{M}(0)$ and $\bar{B}(t)$ are independent of (x,y) the same will be true of $\bar{M}(t,x,y,z)$ and we usually write $\bar{M}(t,z)$.

For this lecture we concentrate on the map $(B_1(t), B_2(t)) \longrightarrow \bar{M}(T,z)$ from the complex valued function of t given by

$$B_1(t) + \sqrt{-1}\, B_2(t)$$

into the vector valued function of z given by $\bar{M}(T,z)$. This can be made more symmetric by replacing $\bar{M}(t,x,y,z)$ by its stereographic projection from $\begin{pmatrix} 1 \\ 0 \\ 0 \end{pmatrix}$,

$$\phi(t,x,y,z) \equiv \frac{M_z + iM_y}{M_x - 1}$$

(M_z, M_x, M_y denote the components of $\bar{M}(t,x,y,z)$ and depend on (t,x,y,z).)
In this fashion we finally get a map between complex valued functions of $0 < t < T$ and complex valued functions $\phi(T,z)$ of $-\infty < z < \infty$. We have coined the terminology "Bloch transform" for this map [24,25]. In these references we proved that the linear part of this map at zero, i.e., for very small applied r.f. fields, is exactly the Fourier transform, a result well understood by the NMR community at least for "small tip angle", see [26].

A number of important questions remain totally unexplored in regards to this "Bloch transform"; here are a few.

(a) What is its range?

(b) Is it invertible away from zero?

(c) Given a limited set of desired final states, is it possible to obtain appropriate initial states?

The rest of the lecture is devoted to describing some recent results obtained in collaboration with A.Hasenfeld, see [27,28].

It is possible (see [24]) to relate the Bloch equations (in all justice, we are only dealing with the case when the decay terms are missing and the equations should be called the Euler equations) with the Riccati equations and in turn this one to the Schroedinger equation. This connection was noticed by S.Lie more than one hundred years ago.

By exploiting some of the new developments around the Schroedinger equation stemming from its relation with, once again, the Korteweg-de Vries equation, one can come up with some rather startling results that allow for an exploration of some of the properties of the "Bloch transform".

The Korteweg-de Vries equation

$$U_t = U_{xxx} + 6UU_x$$

allows a remarkable class of ***explicit*** solutions: the so-called N-soliton solutions. Among these are some even more special ones satisfying

$$V(x,0) = N(N+1)\,\mathrm{sech}^2(x) \quad .$$

These functions can be written as follows:

$$V(x,0) = -2\,\frac{d^2}{dx^2}\log\tau(x)$$

with

$$\tau(x) = \det\left(\delta_{ij} + c_j^2\,\frac{e^{-(i+j)x}}{i+j}\right), \qquad 1 \leqslant i,j \leqslant N,$$

$$c_j = 2j(-1)^{j-1}\prod_{i\neq j}\frac{i+j}{i-j}$$

and, moreover, if one puts

$$\tau(x,t) = \det\left(\delta_{ij} + c_j^2(t)\,\frac{e^{-(i+j)x}}{i+j}\right), \qquad 1 \leqslant i,j \leqslant N$$

$$c_j(t) = \exp(j^3 t)c_j(0)$$

and finally sets

$$V(x,t) = -2\,\frac{d^2}{dx^2}\log\tau(x,t)$$

one has ***solved*** the KdV equation with initial data $V(x,0)$!!

Now I can explain the results described in [27,28] and confirmed both by numerical simulations, i.e. solving the Bloch equations forward in time as well as by actual experiments, [28], carried out with a .5 T Oxford/IBM Imaging Spectrometer at the Lawrence Berkeley Laboratory by A.Hasenfeld.

A word of warning: watch out for some few exchanges of the variables x and t!

In [27] we pick, for simplicity N=2. For higher values, see [28]. For a certain value of t, say t_o, we consider $V(x,t_o)$ as given above as the input in the Bloch equations. We solve numerically these equations with applied r.f. field given by

$(V(\cdot,t_0),\ 0)$ and, as usual, initial condition $\begin{pmatrix}0\\0\\-1\end{pmatrix}$. We run the computation as long as it takes the vector $\bar{M}(t,x,y,0)$, i.e., in resonance, to return to the position $\begin{pmatrix}0\\0\\-1\end{pmatrix}$ for the first time, i.e., to perform a 2π flip. We display the value of $-M_z$ as a function of z. Then we pick another value of t, say t_1, use $V(\cdot,t_1)$ as the input in the Bloch equations for as long as it takes for a 2π flip–on resonance, and once again display the value of $-M_z$. The results are quite interesting. Using $V(x,0)$ we get that all spins return to the initial position, not only those in $z=0$. For values of $t_0 \neq 0$ we get "inversion" of the spins for two bands of the z axis whose widths depend on the value of t_0, i.e. of the input r.f. field.

If we do the complete set of experiments, one for each value of t_0, all over again but waiting for a 4π flip, then we get a ***complete*** return to equilibrium independent of z and for every value of t_0.

A final run with a longer running time to obtain a 6π flip angle – on resonance – gives a picture similar to the 2π flip case, except that the inversion bands are much narrower and closer to the "on resonance" region $z=0$.

The figures on the following three pages illustrate the situation described above.

We observe that the pure N-soliton potentials given by $V(x) = -2\dfrac{d^2}{dx^2}\log\tau(x)$ with

$$\tau(x) = \det\left(\delta_{ij} + c_j^2\,\frac{e^{(-\kappa_i+\kappa_j)x}}{\kappa_i+\kappa_j}\right), \qquad 1 \leqslant i,j \leqslant N$$

can be obtained from $V=0$ by repeated applications of the Darboux process with ***shifts***.

A final note about the KdV equation: if one puts $L = -D^2 + V$ and takes

$$B = ((-L)^{\frac{1}{2}})^3_+ \equiv \text{the differential operator}$$

part of the formal expression

$$(-L)^{3/2} = D^3 + a_2D^2 + a_1D + a_0 + a^{-1}D_{-1} + \dots.$$

we can put the KdV equations in so-called Lax's form

$$\dot{L} = [B,L] \quad .$$

This same form will appear in the next lecture in connection with Toda's equation. Notice, finally, that if we pick

$$B = (-L)^{\frac{1}{2}}_+$$

we get the equation

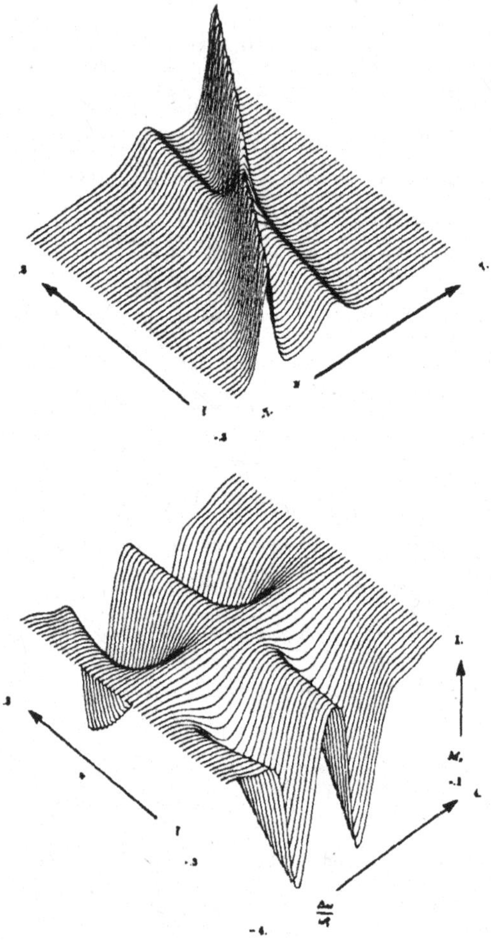

Each of the 50 wave profiles in the two soliton solution $V(\bar{x})$, $\bar{t}$ = constant in the top plot is used as an amplitude modulated 2π pulse in the bottom plot, where the z component of the magnetization is displayed as a function of $\Delta\omega/\omega_1^0$ after the pulse.

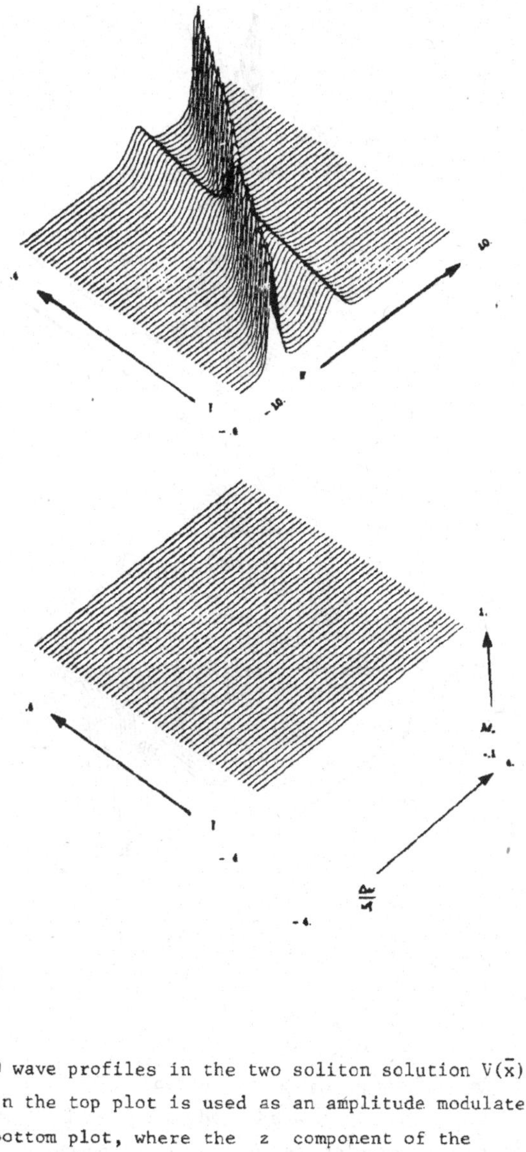

Each of the 50 wave profiles in the two soliton solution $V(\bar{x})$, $\bar{t}$ = constant in the top plot is used as an amplitude modulated 4π pulse in the bottom plot, where the z component of the magnetization is displayed as a function of $\Delta\omega/\omega_1^0$ after the pulse.

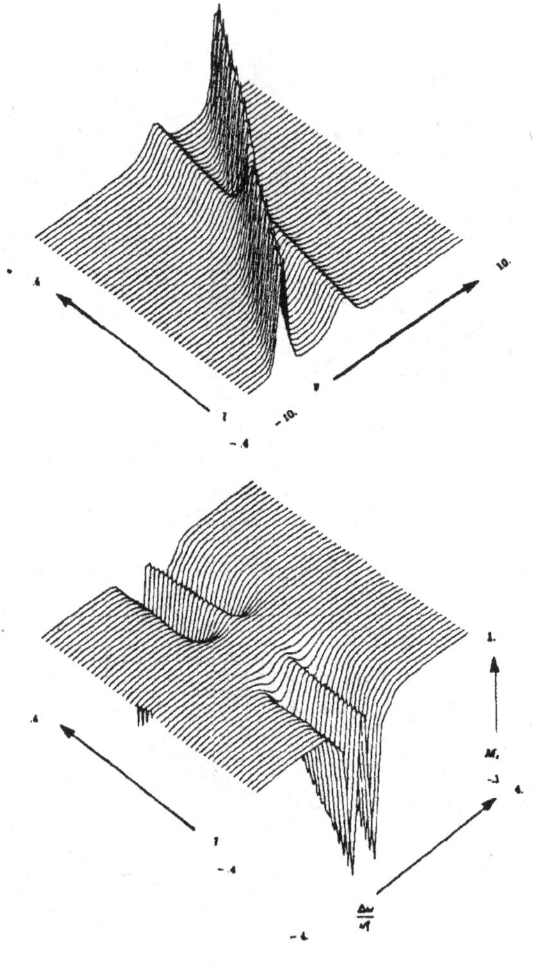

Each of the 50 wave profiles in the two soliton solution $V(\bar{x})$, $\bar{t}$ = constant in the top plot is used as an amplitude modulated 6π pulse in the bottom plot, where the z component of the magnetization is displayed as a function of $\Delta\omega/\omega_1^0$ after the pulse.

$$U_t = U_x$$

giving rise to translations and thus to the Fourier transform.

LECTURE 5: The Toda Lattice

In the last two lectures we have seen, at least a bit of, the role that a nonlinear evolution equation like that of Korteweg-de Vries has played so far in connection with areas rather *remote* from the place of origin.

The same can be said of the Darboux method, and to some extent of the whole circle of ideas originating in the work of Slepian, Landau and Pollak.

Here I give a brief account of another important equation, the "Toda lattice", which has also reared its head in unexpected places. Maybe this effect can be "amplified" by making this material better known.

Recall that a harmonic lattice given by

$$\ddot{x}_n(t) = a_n x_{n+1}(t) + b_n x_n(t) + c_n x_{n-1}(t) \quad , \qquad 1 \leqslant n \leqslant N$$

with a symmetric matrix on the rhs, and appropriate boundary conditions, has normal modes, i.e., any initial vector $x_n(0)$ can be expanded

$$x_n(0) = \sum_{j=1}^{N} V_n^j$$

and

$$x_n(t) = \sum_{j=1}^{N} V_n^j e^{i\lambda_j t} \quad .$$

The energy stored at t=0 in the jth mode V_n^j stays in that mode for all time. A very different story takes place when nonlinear interactions are allowed and one expects "equidistribution of energy" as time goes on. The advent of modern computers ~1945 made it possible for Fermi, Pasta and Ulam to play with this for simple nonlinear lattices. The results were not always as predicted by the folklore: sometimes all the energy seemed to come back to the mode where it was stored originally in a periodic fashion. This was taken up again by M.Kruskal and N.Zabusky around 1965, except that they considered a continuous limit of these spatially discrete equations which gave them, of all things, the KdV equation. The rest, as they say in bad movies, is history.

This attracted a lot of attention. Among the most distinguished pieces of work at that time is that of Toda.

A brief description of this problem, its solution, and some fascinating connections is the purpose of this talk.

Put

$$\dot{Q}_n = P_n$$
$$\dot{P}_n = e^{-(Q_n - Q_{n-1})} - e^{-(Q_{n+1} - Q_n)}$$

with $Q_o = -\infty$, $P_o = 0$,

$Q_{N+1} = +\infty$, $P_{N+1} = 0$, i.e., fixed end masses.

Put, with Flaschka [29,30]

$$a_n = \tfrac{1}{2} e^{-(Q_{n+1} - Q_n)/2} , \qquad b_n = P_n/2$$

to get

$$\dot{a}_n = a_n(b_n - b_{n+1}) ,$$
$$\dot{b}_n = 2(a_{n-1}^2 - a_n^2) .$$

Follow Flaschka one more step in this enchanted trip and set

$$L = \begin{bmatrix} b_1 & a_1 & & \\ a_1 & b_2 & a_2 & \\ & \ddots & \ddots & a_{N-1} \\ & & a_{N-1} & b_N \end{bmatrix} , \qquad B = \begin{bmatrix} 0 & -a_1 & & \\ a_1 & 0 & & \\ & \ddots & \ddots & -a_{N-1} \\ & & a_{N-1} & 0 \end{bmatrix}$$

and observe that the equations of motion can be written in Lax's form

$$\dot{L} = [B,L] \quad (\equiv BL - LB) \quad (t \text{ dependence is implicit}) .$$

Since B is skew hermitian one easily shows that if $U(t)$ denotes the solution of

$$\dot{U}(t) = B(t)U(t) , \qquad U(0) = I ,$$

then $U(t)$ is orthogonal and

$$\frac{d}{dt}(U^{-1}(t)L(t)U(t)) = 0$$

which means

$$L(t) = U(t)L(0)U^{-1}(t) ,$$

i.e., we have an "isospectral evolution": the eigenvalues of $L(t)$ are the same for all $t \geq 0$.

Now a three lines review of tridiagonal matrices: Take L as above, with $a_i > 0$, and denote by $Y_{\cdot,k}$ the eigenvector with eigenvalue λ_k and $\|Y_{\cdot,k}\| = 1$,

$$LY_{\cdot,k} = \lambda_k Y_{\cdot,k} \quad .$$

Since we have

$$b_i = \sum_{k=1}^{N} Y_{ik}^2 \lambda_k$$

and

$$a_{i-1}Y_{i-1,k} + b_i Y_{i,k} + a_i Y_{i+1,k} = \lambda_k Y_{i,k}$$

one shows easily that

$$a_i^2 = \sum_{k=1}^{N} [(\lambda_k - b_i)Y_{i,k} - a_{i-1}Y_{i-1,k}]^2$$

and

$$Y_{i+1,k} = \frac{1}{a_i}((\lambda_k - b_i)Y_{i,k} - a_{i-1}Y_{i-1,k}) \quad .$$

One can use these formulas recursively to show that $\{\lambda_1,\ldots,\lambda_k,\ldots,\lambda_N\}$ and $Y_{1,\cdot}$ determine $b_1, a_1, Y_{2,\cdot}, b_2, a_2, Y_{3,\cdot}, b_3, a_3, Y_{4,\cdot}, \ldots$ in this order. L is thus determined completely.

Now we find an equation for the quantities $Y_{1,k}(t)$. From

$$L(t)Y_{\cdot,k}(t) = \lambda_k Y_{\cdot,k}(t)$$

we get by differentiating and using $\dot{L} = BL-LB$ and $\dot{\lambda}_k = 0$,

$$BLY_{\cdot,k} - LBY_{\cdot,k} + L\dot{Y}_{\cdot,k} = \lambda_k \dot{Y}_{\cdot,k}$$

and using $LY_{\cdot,k} = \lambda_k Y_{\cdot,k}$ one can rewrite this as

$$(L - \lambda_k I)(\dot{Y}_{\cdot,k} - BY_{\cdot,k}) = 0 \quad .$$

Since L has simple spectrum, one gets $\dot{Y}_{\cdot,k} - BY_{\cdot,k} = \alpha_k Y_{\cdot,k}$ but since the length of $Y_{\cdot,k}$ is one, we get $\alpha_k = 0$ and

$$\dot{Y}_{\cdot,k}(t) = B(t)Y_{\cdot,k}(t) \quad .$$

Recall the form of $B(t)$ to get, for the first component

$$\dot{Y}_{1,k}(t) = -a_1(t)Y_{2,k}(t) \quad .$$

On the other hand, $LY_{\cdot,k} = \lambda_k Y_{\cdot,k}$ gives

$$b_1 Y_{1,k} + a_1 Y_{2,k} = \lambda_k Y_{1,k}$$

and we had $b_1 = \sum_{k=1}^{N} \lambda_k Y_{1k}^2$ so we get

$$\left(\sum_{k=1}^{N} \lambda_k Y_{1,k}^2(t) - \lambda_k\right) Y_{1,k}(t) = \dot{Y}_{1,k}(t) .$$

This is "solved" by

$$Y_{1,k}(t) = Y_{1,k}(0)\, e^{-\lambda_k t + \int_0^t \Sigma \lambda_k Y_{1,k}^2(s) ds} .$$

Square this, multiply by λ_k, and then sum, to get

$$\sum_{k=1}^{N} \lambda_k Y_{1,k}^2(t) = \Sigma \lambda_k Y_{1,k}^2(0)\, e^{-2\lambda_k t + 2\int_0^t \Sigma \lambda_k Y_{1,k}^2(s) ds}$$

or setting $h(t) \equiv \sum_1^N \lambda_k Y_{1,k}^2(s) ds$,

$$-2h(t)\, e^{-2\int_0^t h(s)ds} = -2\Sigma \lambda_k Y_{1,k}^2(0)\, e^{-2\lambda_k t}$$

which integrates to

$$e^{-2\int_0^t h(s)ds} = \sum_1^N e^{-2\lambda_k t}\, Y_{1,k}^2(0) .$$

Returning to the expression for $Y_{1,k}(t)$ we get

$$Y_{1,k}(t) = \frac{Y_{1,k}(0)\, e^{-\lambda_k t}}{\sqrt{\Sigma e^{-2\lambda_k t}\, Y_{1,k}^2(0)}} .$$

Now we can put this to use. With $\lambda_1 < \lambda_2 < \ldots < \lambda_N$, put $\lambda = \min(\lambda_{k+1} - \lambda_k)$. Then

$$1 - Y_{1,1}^2(t) = 1 - \frac{Y_{1,1}^2(0)\, e^{-2\lambda_1 t}}{\Sigma e^{-2\lambda_k t}\, Y_{1,k}^2(0)} = 1 - \frac{1}{1 + \sum_{k>1} e^{-2(\lambda_k - \lambda_1)t} \left(\frac{Y_{1,k}}{Y_{1,1}}\right)^2}$$

and expanding we see that this is bounded by an expression of the form

$$\mathbb{C}\, e^{-2\lambda t}$$

since

$$b_1 - \lambda_1 = \sum_{k=1}^{N} (\lambda_k - \lambda_1) Y_{1,k}^2 = \sum_{k>1} (\lambda_k - \lambda_1) Y_{1,k}^2$$

and

$$\sum_{k>1} Y_{1,k}^2 + Y_{11}^2 = 1 \implies \sum_{k>1} Y_{1,k}^2 < \mathbb{C}\, e^{-2\lambda t}$$

we get $|b_1 - \lambda_1| \leq D\, e^{-2\lambda t}$. Finally we have

$$a_1^2 = \sum_{k=1}^{N} (\lambda_k - b_1)^2\, Y_{1,k}^2$$

and here all terms die exponentially, so that

$$a_1^2 \leq K e^{-2\lambda t} .$$

The same argument can now be pushed by induction for larger values of the index, and one gets that as $t \to \infty$,

$$\lim a_i(t) = 0$$

$$\lim b_i(t) = \lambda_i$$

both at an exponential rate governed by λ. This means that by solving the equations $\dot{L} = [B,L]$ forward in time, one could in principle determine the eigenvalues of $L(0)$. Indeed, the beautiful development reported in [31] shows that a properly constructed isospectral evolution in the space of tridiagonal matrices has the property that when sampled at integer values of t it coincides with the Q-R algorithm. We refer the reader to [31] for this surprising result.

Other isospectral evolutions are obtained by replacing

$$\dot{L} = [L^+ - L^-, L]$$

by

$$\dot{L} = [(L^p)_+ - (L^p)_-, L]$$

where L_+ and L_- denote the strictly upper and lower triangular parts of a matrix L.

LECTURE 6: Tomographic Methods in Radar Detection

Here we show how the problem of range-Doppler radar detection can be interpreted within the context of the "Radon transform". The exposition here is an introduction to a fuller treatment given in [32]. The problem is that of determining an unknown distribution of objects in space with $D(r,v)$ the joint distribution for range and velocity.

We denote with $s(t)$ the transmitted pulse $s(t)$ and with $\psi(t)$ its complex valued waveform

$$\psi(t) = s(t) + \frac{i}{\pi} \text{P.V.} \int_{-\infty}^{\infty} \frac{s(\tau)}{t-\tau} d\tau \quad .$$

We will consider pulses of the form

$$\psi(t) = U(t)\, e^{2\pi i f_0 t}$$

with the carrier frequency f_0 much larger than the bandwidth of $U(t)$.

An object at range r, moving with velocity v towards the detector-source antenna, produces an echo $e(t)$ whose waveform $\psi e(t)$ is

$$\psi e(t) = D\psi(t-x)\, e^{-2\pi i y t} \quad , \qquad x = \frac{2r}{c} \quad , \qquad y = 2f_0 v/c \quad .$$

Going to the variables x,y the effect of a collection of objects distributed according to $D(x,y)$ is

$$\psi e(t) = \iint D(x,y)\psi(t-x)\, e^{-2\pi i y t}\, dxdy \quad .$$

We will see here that by forming "matched filter" integrals of the type

$$\int_{-\infty}^{\infty} \psi e(t)\, \bar{\psi}_{\tau,\nu}(t) dt$$

with

$$\psi_{\tau,\nu}(t) \equiv \psi(t-\tau)\, e^{-2\pi i \nu t}$$

one can reconstruct the function $D(x,y)$. This will be accompished by choosing appropriately the "pulse shapes" $\psi(t)$.

The integral above can be expressed as

$$A(\tau,\nu) = \iint D(x,y)\, e^{\pi i(\nu-y)(x+\tau)}\, A_\psi(x-\tau,\, y-\nu) dxdy$$

with the so-called "ambiguity function of ψ" given by

$$A_\psi(s_1,s_2) \equiv \int_{-\infty}^{\infty} \psi(t - \tfrac{1}{2}s_1)\bar{\psi}(t + \tfrac{1}{2}s_1)\, e^{-2\pi i s_2 t}\, dt \quad .$$

This function holds the key to the problem. It has appeared in several setups either in this form or in related forms, and is very much related to the "Wigner function". Let me observe that at times one only measures the ***modulus*** of the "matched filter" given above. On a related vein one can investigate the problem of recovering ψ from A_ψ (this can be done) or from $|A_\psi|$ (this cannot be done) (see [33]).

The most relevant properties of $A_\psi(s_1,s_2)$ in our problem are

1. If $\psi_\sigma(t) = (\pi\sigma)^{-1/4}\, e^{-t^2/2\sigma}\, e^{2\pi i f_0 t}$, $\sigma > 0$, then

$$A_\psi(s_1,s_2) = e^{-2\pi i f_0 s_1}\, e^{-\tau^2/4\sigma}\, e^{-\pi^2 s_2^2 \sigma} .$$

2. Introduce the "chirp" of a complex wave form $\psi(t)$,

$$\theta_\mu(t) \equiv \mu^{-1/2} \int_{-\infty}^{\infty} e^{-\pi i\omega^2/\mu}\, \psi(\omega-t)\,d\omega , \qquad \mu \neq 0,$$

then one proves easily that

$$A_{\theta_\mu}(s_1,s_2) = A_\psi(s_1 - \mu s_2, s_2).$$

Combining these two results we see that for any $\mu \neq 0$ one can take σ small enough and get a wave form $\theta(\mu,\sigma)$ such that

$$\tfrac{1}{2}\int_{-\infty}^{\infty}\int_{-\infty}^{\infty} F(s_1,s_2) A_{\theta(\mu,\sigma)}(s_1,s_2)\,ds_1\,ds_2 \cong \int_{-\infty}^{\infty} F(\mu s_2, s_2)\,ds_2$$

for any reasonable function $F(s_1,s_2)$.

Going back to the expression $A(\tau,\gamma)$ obtained earlier for the "matched filter" we see that by choosing σ small we can approximate the integral of the function

$$D(x,y)\, e^{\pi i(\nu-y)(x+\tau)}$$

along any line through the origin with equation

$$x-\tau = \mu(y-\nu) .$$

Substituting this in the double integral expressing $A(\tau,\nu)$ we get

$$\int_{-\infty}^{\infty} D(\tau + \mu y - \mu\nu,\ y)\, e^{\pi i(\nu-y)(2\tau + \mu y - \mu\nu)}\, dy .$$

Now with the substitutions $y \longrightarrow y+\nu - \tau/\mu$ and $\eta \equiv \gamma - \tau/\mu$ we can express this as

$$e^{-\pi i\tau^2/\mu} \int_{-\infty}^{\infty} D(\mu y,\ y+\eta)\, e^{-\pi i\mu y^2}\, dy .$$

Thus we have reduced the problem to a version of the Radon inversion problem. The main twist here is the presence of the exponential factor in the integral. By varying

μ, η we can scan all lines in the plane. Restricting μ gives a version of the "limited angle" problem in tomography.

As to uniqueness of a solution, one can prove it using that D is real valued and supported in $x \geq 0$. See [32] for details and numerical simulations. As remarked in [32], one can interpret our procedure of varying μ to obtain different wave forms to "probe" D(x,y) as similar to that used in magnetic resonance imaging. In both of these cases one replaces mechanical motion by electronic encoding of spatial information. In NMR this is achieved by playing around with gradients, and here by chirping or linear frequency modulation.

REFERENCES

[1] Bockwinkel, "On the propagation of light in a two-axial crystal around a center of vibration," Versl. Kon Akad. v. Wet. XIV 2 (1906), 636-651.

[2] Radon,J., "Uber die Bestimmung von Funktionen durch ihre Integral werte längs gewisser Mannifaltigkeiten," Berichte Saechsische Akademic der Wissenschaften 69 (1917), 262-277.

[3] Herman,G.T., *Image Reconstruction from Projections*, (Academic Press, New York, 1980).

[4] Herman,G.T., editor, Proceedings of the IEEE, Special Issue on Computerized Tomography, vol. 71, no. 3 (March 1983).

[5] Davison,M.E. and F.A.Grünbaum, "Tomographic reconstruction with arbitrary directions," Communications on Pure and Applied Mathematics 34 (1981), 77-120.

[6] Smith,K.T., D.C.Solmon and S.L.Wagner, "Practical and mathematical aspects of the problem of reconstructing objects from radiographs," Bulletin of the American Mathematical Society, 83 (1977), 1227-1270.

[7] Gerchberg,R.W. and W.O.Saxton, "A practical algorithm for the determination of phase from image and diffraction plane pictures," Optik, 35, no. 2 (1972), 237-246.

[8] Tuy.H and A.Lent, "An iterative method for the extrpolation of band-limited functions," J. Math. Analysis & Appl. 83, 2, (1981), 554-565.

[9] A.Papoulis, "A new algorithm in spectral analysis and band limited extrapolation," IEEE Trans. on Circuits and Systems, vol. CAS-22, no. 9 (1975), 735-742.

[10] Slepian,D. and H.P.Pollak, "Prolate spheroidal wave functions, Fourier analysis and uncertainty, I." Bell Systems Technical Journal, 40, no. 1 (1961), 43-64.

[11] Landau,H.J. and Pollak,H.P., "Prolate spheroidal wave functions, Fourier analysis and uncertainty, II." Bell Systems Technical Journal, 40, no. 1 (1961), 65-84.

[12] Landau,H.J. and Pollak,H.P., "Prolate spheroidal wave functions, Fourier analysis and uncertainty, III." Bell Systems Technical Journal, 41, no. 4 (1962), 1295-1336.

[13] Slepian,D., "Prolate spheroidal wave functions, Fourier analysis and uncertainty, IV," Bell Systems Technical Journal, 43, no. 6 (1964), 3009-3058.

[14] Slepian,D., "Prolate spheroidal wave functions, Fourier analysis and uncertainty," Bell Systems Technical Journal, 57, no. 5 (1978), 1430-2371.

[15] Slepian,D., "Some comments on Fourier analysis, uncertainty and modelling," SIAM Review, 25 (1983), 379-394.

[16] Grünbaum,F.A., "A new property for reproducing kernels of classical orthogonal polynomials," J. of Math. Analysis and Appl., 95 (1983), 491-500.

[17] Perlstadt,M., "Chopped orthogonal polynomial expansions – some discrete cases," SIAM J. Alg. Disc. Meth. 4 (1), (1983), 94-100.

[18] Perlstadt,M., "A property of orthogonal polynomial families with polynomial duals," SIAM J. Math. Anal. 15 (5), (1984), 1043-1054.

[19] Perline,R., "Discrete time-band limiting operators and commuting tridiagonal matrices," to appear in SIAM J. Alge. Disc. Meth.

[20] Duistermaat,J. and F.A.Grünbaum, "Differential equations in the eigenvalue parameter," Comm. Math. Physics, 103 (1986), 177-240.

[21] Grünbaum,F.A., "A remark on Hilbert's matrix," Linear Algebra and its Applications, 43 (1982), 119-124.

[22] Bertero,M. and F.A.Grünbaum, "Commuting differential operators for the finite Laplace transform," Inverse Problems, 1 (1985), 181-192.

[23] Coddington,E. and N.Levinson, *Theory of Ordinary Differential Equations*, (McGraw-Hill Book Co., 1955).

[24] Grünbaum,F.A., "The Bloch equations and two nonlinear extensions of the Fourier transform," Proceedings, Third Annual Meeting, Society of Magnetic Resonance in Medicine, New York (1984), 281-282.

[25] Grünbaum,F.A., "An inverse problem related to the Bloch equations and the nonlinear Fourier transform," Inverse Problems, 1 (1985), L25-L28.

[26] Hoult,D., "The solution of the Bloch equations in the presence of a varying B_1 field," J. Magnetic Resonance, 35 (1979), 69.

[27] Grünbaum,F.A. and Hasenfeld,A., "An exploration of the invertibility of the Bloch transform," Inverse Problems, 2 (1986), 75-81.

[28] Hasenfeld,A., "An inverse problem for in vivo NMR spatial localization," Ph.D. Thesis, University of California, Berkeley (1985).

[29] Flaschka,H., "The Toda lattice, I: Existence of integrals," Phys. Rev. B., 9, (1974), 1924-1925.

[30] Flaschka,H., "On the Toda lattice, II: Inverse scattering solution," Prog. Theor. Physics, 51 (1974), 703-716.

[31] Deift,P., T.Nanda and C.Tomei, "Ordinary differential equations and the symmetric eigenvalue problem," SIAM J. Numer. Analysis, 20, 1 (1983), 1-22.

[32] Feig,E. and F.A.Grünbaum, "Tomographic methods in range-Doppler radar," to appear in Inverse Problems, vol. 2, 2 (1986).

[33] Grünbaum,F.A., "A remark on the radar ambiguity function," Transactions on Information Theory, IEEE, vol. IT-30, no. 1, (January 1984), 126-127.

NUMERICAL TREATMENT OF ILL-POSED PROBLEMS

Frank Natterer

1. INTRODUCTION

A problem is said to be ill-posed if its solutions is either not well defined (i.e. it does not exist or it is not uniquely determined) or it does not depend continuously on the data. The theory of ill-posed problems is covered in a series of monographs (Tikhonov-Arsenin (77), Morozov (84), Lavrentiev et al. (83), Groetsch (84), Bertero (82)) and will not be dealt with here. Rather we concentrate on the numerical techniques which are being used after a linear ill-posed problems has been reduced to a problem of linear algebra by some discretization procedure. We do not describe this discretization process in detail (a selection of papers dealing with the various aspects of discretization is Nashed (76), Natterer (77), (83), Engl (83), Ivanov (76), see also Groetsch (84)). For the first-kind integral equation

$$\int_a^b K(x,y)f(y)dy = g(x) \qquad , \quad c \leq x \leq d$$

which is a typical instance of an ill-posed problem, an obvious discretization is obtained by a quadrature rule with weights w_j, turning the integral equation into

$$\sum_{j=1}^{m} w_j K(x_i,y_j) f(y_j) = g(x_i) \qquad , \quad i = 1,\ldots,n \quad .$$

More sophisticated discretizations use an n-dimensional subspace $V_n = sp\{v_1,\ldots,v_n\}$ and m linear independent functionals $\psi_1,\ldots,\psi_m$ to define an approximate solution f_n of the operator equation $Kf = g$ by

$$\psi_i(Kf - g) = 0 \qquad , \quad f \in V_n \qquad .$$

In any case we end up with a linear system

$$Ax = b \tag{1.1}$$

where A is a (m,n) - matrix and $x \in \mathbb{R}^n$, $b \in \mathbb{R}^m$. The numerical solution of this system is the subject of the paper. We shall outline a few problems related to (1.1), and we shall describe in some detail numerical algorithms for coping with them.

The linear system (1.1) may be underdetermined or overdetermined. In order to restore unique solvability of (1.1) we introduce the notion of a generalized solution: Among all vectors x minimizing the residual $\|Ax-b\|$, we select the one with least norm $\|x\|$ and denote it by A^+b. Here and in the following we use the euclidean norm. Then, A^+ is a well defined (n,m) - matrix which is called the (Moore-Penrose) generalized inverse of A. Clearly, $A^+ = A^{-1}$ if A is an invertible (n,n)-matrix. It is easily seen that $x = A^+g$ is the unique solution of the normal equations

$$A^*Ax = A^*g \qquad , \quad A^* \text{ the transpose of } A \ , \tag{1.2}$$

which is in the range $R(A^*)$ of A^*. See Ben-Israel et al. (74) and Nashed (76) for theory and application of generalized inverses. While the use of A^+ instead of A^{-1} disposes of possible non-existence and non-uniqueness, it does not alleviate the problem of instability. Restoring stability is the most interesting and also the most difficult part in dealing with an ill-posed problem. There are several regularization methods for restoring stability. The most widely used of these methods is probably the Tikhonov-Phillips method which consists in defining a regularized solution $x = R_\gamma b$ of $Ax = b$ by minimizing

$$\|Ax-b\|^2 + \gamma^2 \|x\|^2 \quad . \tag{1.3}$$

The positive number γ is called regularization parameter. Obviously, $x = R_\gamma b$ is the unique solution of the regularized normal system

$$(A^*A + \gamma^2 I)x = A^*b \quad . \tag{1.4}$$

Sometimes x is known in advance to lie in a set $M \subseteq \mathbb{R}^n$. If this is the case it appears resonable to minimize (1.3) in M rather than in all of $\mathbb{R}^n$. We then have to solve a constrainedquadratic optimization problem. Typical constraints are non-negativity, monotonicity and convexity of x.

An other regularization method makes use of the singular value decomposition (SVD) of the matrix A. Let $\sigma_1^2,\ldots,\sigma_p^2$ be the positive eigenvalues of the symmetric matrix A^*A, and let $v_1,\ldots,v_p$ be the corresponding orthonormal system of eigenvectors, i.e.

$$A^*Av_k = \sigma_k^2 v_k \quad , \quad (v_k,v_\ell) = \delta_{k\ell} \quad . \tag{1.5a}$$

Let $u_k = \sigma_k^{-1}Av_k$. It follows readily that

$$AA^*u_k = \sigma_k^2 u_k \quad , \quad (u_k,u_\ell) = \delta_{k\ell} \quad . \tag{1.5b}$$

Obviously, each Ax in $R(A)$ admits an expansion in terms of the $u_1,\ldots,u_p$ which is readily seen to be

$$Ax = \sum_{k=1}^{p} \sigma_k (x,v_k)u_k \quad .$$

In matrix notation, (1.5) reads

$$A = U \Sigma V^* \tag{1.6}$$

where U, V are orthonormal and Σ is diagonal. More explicitly,

$$U = (u_1,\ldots,u_p) \quad , \quad \Sigma = \begin{pmatrix} \sigma_1 & & \\ & \ddots & \\ & & \sigma_p \end{pmatrix} \quad , \quad V = (v_1,\ldots,v_p) \quad .$$

The representation (1.6) of A is called the singular value decomposition (SVD) of A. The numbers $\sigma_1 \geq \ldots \geq \sigma_p > 0$ are called the singular values of A. A^+ can be written as

$$A^+ = V \Sigma^{-1} U^* \quad ,$$

$$A^+ b = \sum_{k=1}^{p} \sigma_k^{-1} (b, u_k) v_k \quad .$$

We see that instabilities in the evaluation of $A^+ b$ occur if and only if some of the singular values are much smaller than σ_1. For such singular values, $\sigma_k^{-1}(b,u_k)v_k$ cannot be computed accurately. A stable approximation to $A^+ b$ is given by

$$R_\gamma b = \sum_{\sigma_k \geq \gamma} \sigma_k^{-1} (b, u_k) v_k \quad . \tag{1.7}$$

This is known as the truncated SVD of A^+. More generally, we can put

$$R_\gamma b = \sum_{k=1}^{p} F\left(\frac{\sigma_k}{\gamma}\right) \sigma_k^{-1} (b, u_k) v_k \tag{1.8}$$

with a function $F(\sigma)$ which is close to 0 for σ small and which tends to 1 as $\sigma \to \infty$. In this context, F is called a filter, and (1.8) is referred to as digital filtering. Examples for F are

$$F(\sigma) = \begin{cases} 0 & , \quad 0 \leq \sigma \leq 1 \\ 1 & , \quad \sigma > 1 \end{cases} \quad ,$$

$$F(\sigma) = \frac{1}{1 + \sigma^{-2}} \quad ,$$

the former one giving truncated SVD, the latter one leading to the Tikhonov-Phillips regularized solution, as is easily seen from (1.4).

The regularization methods described so far are very well established and are widely used in practice. As an example of a less well-known regularization technique we mention the maximal entropy method, see Smith-Grandy (85). Assume $Ax = b$ to be underdetermined. Among all solutions to that equation we then select the one for which the entropy

$$\sum_{j=1}^{n} x_j \log x_j$$

is maximal. Clearly this is similar to the generalized inverse approach, with $\|x\|^2$ replaced by the entropy.

From this tour d'horizon it should be clear that a numerical analyst's toolkit for ill-posed problems should contain routines for computing generalized solutions and regularized solutions

computing the SVD

minimizing a quadratic form subject to constraints

computing maximum entropy solutions.

In the following sections we study the numerical techniques which are presently being used from a computational point of view. For surveys on this subject the reader is referred to Varah (79) and Björck - Elden (79).

2. PERTURBATION THEORY

Since any numerical computation is contaminated by round-off errors we start our study of algorithms with an error analysis.

For an (m,n)-matrix A we put as usual

$$\|A\| = \sup_{x \neq 0} \|Ax\| / \|x\|$$

where $\|x\|$ is the euclidean norm of x. From the SVD of A we see that $\|A\| = \sigma_1$, $\|A^+\| = \sigma_p^{-1}$. The condition number of A is defined to be

$$\kappa(A) = \|A\| \; \|A^+\| = \sigma_1 / \sigma_p \quad .$$

The condition number plays an important role in the numerical solution of linear systems, as can be seen from the following standard result of numerical analysis (see e.g. Isaacson - Keller (66), sec. 1.3):

<u>LEMMA 2.1:</u> Let $Ax = b$, A invertible, and $\tilde{A}\tilde{x} = \tilde{b}$ linear systems with (n,n) - matrices A, $\tilde{A}$. Let $\Theta = \kappa(A)\, \|A-\tilde{A}\| / \|A\| < 1$. Then, the second system is uniquely solvable, and

$$\frac{\|x-\tilde{x}\|}{\|x\|} \leq \frac{\kappa(A)}{1-\Theta} \left\{ \frac{\|A-\tilde{A}\|}{\|A\|} + \frac{\|b-\tilde{b}\|}{\|b\|} \right\} \quad .$$

The lemma says that the (relative) error of A, b enters the solution x amplified by a factor which is essentially $\kappa(A)$.

In the next theorem we extend this result to the generalized solution of overdetermined systems, see also Björck (67), Golub - Saunders (70), Wedin (73), and Lawson - Hanson (74).

THEOREM 2.2: Let $m \geq n$, and let A have full rank. Let x be the generalized solution of $Ax = b$, and let $\tilde{x}$ be the generalized solution of the perturbed system $\tilde{A}\tilde{x} = \tilde{b}$. Then,

$$\frac{\|x-\tilde{x}\|}{\|x\|} \leq \frac{\sqrt{2}+1}{\sqrt{2}-1} \frac{\kappa(A)}{1-\Theta} \left\{ \frac{\|A-\tilde{A}\|}{\|A\|} + \frac{\|b-\tilde{b}\|}{\|b\|}\right\} \left\{1 + \left(\frac{\kappa(A)}{2} \frac{\|b-Ax\|}{\|Ax\|}\right)^2\right\}^{1/2} \tag{2.1}$$

provided that

$$\Theta = \frac{\sqrt{2}+1}{\sqrt{2}-1} \frac{\|A-\tilde{A}\|}{\|A\|} \kappa(A) < 1 \quad .$$

PROOF: In our case, A has precisely n singular values $\sigma_1, \ldots, \sigma_n$, and the normal equations (1.2) are uniquely solvable. Putting $r = (b-Ax) / 2\sigma_n$, these can be written as $A^* r = 0$. Hence x is characterized by

$$C\begin{pmatrix} r \\ x \end{pmatrix} = \begin{pmatrix} b \\ 0 \end{pmatrix} \quad , \quad C = \begin{pmatrix} 2\sigma_n I & A \\ A^* & 0 \end{pmatrix} \quad .$$

C is a symmetric matrix whose eigenvalues are easily seen to be

$$\lambda_{1,k} = \sigma_n + \sqrt{\sigma_n^2 + \sigma_k^2} \quad , \quad \lambda_{2,k} = \sigma_n - \sqrt{\sigma_n^2 + \sigma_k^2} \quad , \; k=1,\ldots,n \; .$$

Thus the largest (smallest) eigenvalues in absolute values of C are

$$\lambda_{1,1} = \sigma_n + \sqrt{\sigma_n^2 + \sigma_1^2} \quad , \quad \lambda_{2,n} = (1 - \sqrt{2})\,\sigma_n$$

and we obtain

$$\|C\| = \lambda_{1,1} \geq \sigma_1 = \|A\| \quad , \quad \kappa(C) = \frac{\lambda_{1,1}}{|\lambda_{2,n}|} \leq \frac{\sqrt{2}+1}{\sqrt{2}-1} \frac{\sigma_1}{\sigma_n} = \frac{\sqrt{2}+1}{\sqrt{2}-1} \kappa(A) \quad .$$

Now define $\tilde{C}$ by replacing A in C by $\tilde{A}$. It follows from Lemma 2.1 that the linear system

$$\tilde{C}\begin{pmatrix} \tilde{r} \\ \tilde{x} \end{pmatrix} = \begin{pmatrix} \tilde{b} \\ 0 \end{pmatrix}$$

is uniquely solvable if $\Theta = \kappa(C)\,\|C-\tilde{C}\| / \|C\| < 1$, and

$$\left\|\begin{pmatrix} \tilde{r}-r \\ \tilde{x}-x \end{pmatrix}\right\| / \left\|\begin{pmatrix} r \\ x \end{pmatrix}\right\| \leq \frac{\kappa(C)}{1-\Theta} \left\{ \|C-\tilde{C}\| / \|C\| + \|b-\tilde{b}\| / \|b\| \right\} \quad .$$

With

$$\|C-\tilde{C}\| = \|A-\tilde{A}\| \quad , \quad \|Ax\| \leq \|b\|$$

we obtain (2.1). □

From (2.1) we conclude that the error in x is essentially $\kappa(A)$ times the error of A, b, provided

$$\kappa(A)\,\|b-Ax\| \,/\, \|b\|$$

is not too large. This is usually the case in ill-posed problems since $\|b-Ax\| / \|b\|$ has the order of magnitude of the measurement error.

Now let us consider the computation of x from the normal equations (1.2). The condition number is now $\kappa(A^*A) = \kappa^2(A)$. This means that perturbations in A^*A, A^*b are amplified by a factor of $\kappa^2(A)$ rather than $\kappa(A)$. Since $\kappa(A)$ is typically a very large number for an ill-posed problem this may cause a severe problem with round-off in the solution process. Assume that the computations are done on a t-digit machine. Then, A^*A, A^*b can be computed with an accuracy of 10^{-t}, and the round-off errors made during the solution process can also be interpreted as perturbations of A^*A, A^*b of this size by a backward analysis, see Wilkinson (63). The accumulated round-off error is then of order $\kappa^2(A)\ 10^{-t}$. This is much larger than the $\kappa(A)\,10^{-t}$ - error which we expect from the above error analysis based on Theorem 2.2. Therefore, computing generalized solutions by putting up and solving the normal equations cannot, in general, be recommended.

It is important to understand that this "squaring of the condition number" is a purely numerical problem. It has nothing to do with the amplification of those errors in A, b which are caused by inaccurate measuring, inadequate modelling, or discretization. If the arithmetic of our machine is sufficiently accurate so as to make $\kappa^2(A)\,10^{-t}$ less than the accuracy we need in the computation of x it is o.k. to use the normal equations. For instance one always can use the normal equations if one is willing to work in double precission.

The same problem occurs in the solution of the regularized normal equations (1.4). Here, the condition number is of the order γ^{-2}, while $\|R_\gamma\|$ is essentially γ^{-1}. There is an other drawback of the regularized normal equations. Since the regularization parameter γ is in general not known in advance, (1.4) has to be solved for several values of γ. If a standard elimination procedure such as Cholesky decomposition is used, then the decomposition has to be recomputed for each new γ, requiring $n^3/6$ operation for each γ.

Now we turn to the error analysis for the SVD, see Lawson - Hanson (74).

The following lemma is a standard result, see Wilkinson (65), sec. II.45.

LEMMA 2.3: Let $A, \tilde{A}$ be symmetric (n,n) - matrices with eigenvalues $\lambda_i, \tilde{\lambda}_i$ resp., $i = 1,\dots,n$, in nonincreasing order. Then,

$$|\lambda_i - \tilde{\lambda}_i| \leq \|A - \tilde{A}\| \qquad , \quad i = 1,\dots,n \qquad .$$

THEOREM 2.4: Let $A, \tilde{A}$ be (m,n) - matrices with singular values $\sigma_1 \geq \dots \geq \sigma_p$, $\tilde{\sigma}_1 \geq \dots \geq \tilde{\sigma}_p$ resp. (some singular values may be missing). Then,

$$|\sigma_i - \tilde{\sigma}_i| \leq \|A - \tilde{A}\| \qquad , \quad i = 1,\dots,p \qquad . \tag{2.2}$$

PROOF: The matrix

$$C = \begin{pmatrix} O & A \\ A^* & O \end{pmatrix}$$

has as eigenvalues the numbers $\pm\sigma$ where σ runs through all the singular values of A, completed by the eigenvalue O of appropriate multiplicity. Define $\tilde{C}$ by replacing A in C by $\tilde{A}$. Applying Lemma 2.3 to $C, \tilde{C}$ gives (2.2). □

From (2.2) we see that the error in the computation of the singular values is essentially $\|A - \tilde{A}\|$ if we work on the disturbed matrix $\tilde{A}$. On a t-digit machine we therefore expect an error of order 10^{-t} in the singular values.

However, if we compute the singular values from (1.5a), we get from Lemma 2.3 the estimate

$$|\sigma_i^2 - \tilde{\sigma}_i^2| \leq \|A^*A - \tilde{A}^*\tilde{A}\| \qquad .$$

The right hand side is of order 10^{-t}. We see that σ_i can not be computed accurately unless $\sigma_i^2 >> 10^{-t}$, i.e. $\sigma_i > 10^{-t/2}$.

In order to demonstrate that this discussion is not purely academic we did some numerical experiments. We computed the singular values of the matrix with elements

$$a_{ij} = e^{-10(i/m-j/n)^2} \quad , \quad i = 1,\dots,m \ , \quad j = 1,\dots,n \ .$$

For $n=8$, $m=12$ we obtained the following singular values:

i	σ_i (Golub - Reinsch)		σ_i ((1.5a))	
1	4.71185	(0)	4.71185	(0)
2	2.98644	(0)	2.98645	(0)
3	1.41213	(0)	1.41213	(0)
4	5.04969	(-1)	5.04972	(-1)
5	1.38113	(-1)	1.38114	(-1)
6	2.88547	(-2)	2.88662	(-2)
7	4.45951	(-3)	5.00614	(-3)
8	4.52008	(-4)	1.77938	(-3)

The second column has been computed by the Golub - Reinsch SVD algorithm (LSVDF of the IMSL library) to be described in section 4, while the third column has been computed via (1.5a) using the IMSL routine EIGRS. The computation have been done in single precission on an IBM/370 for which $t=6$. We see that the small singular values (those which are in the order of magnitude 10^{-3} or less) in the third column deviate considerable from those in column 2. This is in full agreement with our analysis.

Again we point out that the problem with (1.5) is a purely numerical one. If the small singular values are not needed (e.g. if the SVD is truncated anyway), using (1.5) is o.k.

3. ALGORITHMS FOR GENERALIZED AND REGULARIZED SOLUTIONS

In the section we describe numerical methods which do not suffer from the drawbacks of the normal equations. We start with computing the generalized solution of $Ax=b$ in the overdetermined case, i.e. $m \geq n$ and A has rank n. Assume $A=QR$, where Q is a unitary (m,n)-matrix (i.e. $Q^*Q = I_n$ = the n-dimensional unit matrix) and R is upper triangular. Then, the normal equations $A^*Ax = A^*b$ can be written as $R^*Q^*QRx = R^*Q^*b$, and this simplifies into

$$Rx = Q^*b \quad . \tag{3.1}$$

Since $A^*A = R^*R$, A and R have the same singular values, and it follows that $\kappa(R) = \kappa(A)$. Hence solving (3.1) does not suffer from squaring the condition number and can be solved stably and also efficiently by back-substitution. Hence everything depends on a satisfactory way to compute the QR - decomposition of A.

For later use we describe the familiar Householder process to compute $A = QR$. Here, Q is obtained as the first n columns of a product $Q_1 \cdots Q_n$ of symmetric unitary (m,m) - matrices which are determined such that

$$Q_k \cdots Q_1 \; A$$

is upper triangular in the first k columns for $k = 1,\ldots,n$. Q_1 is chosen so as to map the first column a_1 into a multiple $\pm \|a_1\| e_1$ of the first unit vector. This is achieved by putting $Q_1 = I_m - 2u_1u_1^*$ where the vector u_1 of length 1 is given by

$$u_1 = (a_1 \pm \|a_1\| e_1) / \alpha_1$$

with a normalization factor α_1. In order to avoid cancellation the sign $\pm$ is chosen to be the sign of the first element of a_1. Obviously, Q_1A is of the form

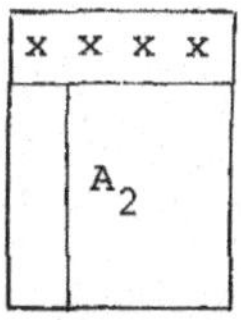

with zeros in the first column below the (1,1) element.

In a second step we choose a matrix Q_2 of the form

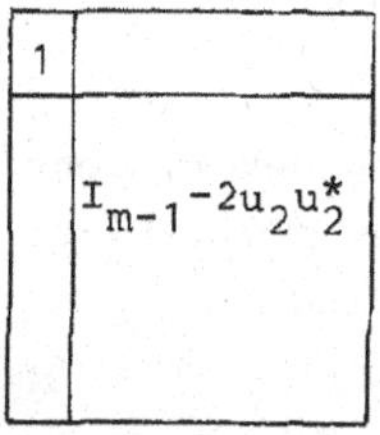

where the vector $u_2 \in \mathbb{R}^{m-1}$ is chosen so as to map the first column of A_2 into a multiple of the first unit vector of $\mathbb{R}^{m-1}$. After n steps we arrive at the desired decomposition. The product of the matrices Q_k is not computed explicitly. Rather we compute the right hand side of (3.1) by successive application of the matrices Q_k, i.e.

$$Q^*b = Q_n \cdots Q_1 b \quad .$$

This is done most easily by simply appending b as $(n+1)$ - st column to the matrix A in the decomposition process.

The operation count for the whole Householder process is $\frac{2}{3}n^3 + \mathcal{O}(n^2)$ for $m = n$. This coincides with the number of operations we need for putting up the normal equations $(\frac{1}{2}n^3)$, followed by a Cholesky decomposition $(\frac{1}{6}n^3 + \mathcal{O}(n^2))$. However, an analysis of the round-off error (see Golub (65), Lawson - Hanson (74)) reveals that the Householder process, due to the orthogonality of the matrices Q_k, is much more stable than factoring the normal equations.

In the underdetermined case, i.e. $m \leq n$ and A has rank m, the generalized solution of $Ax = b$ is the unique solution of $A^*Ax = A^*b$ in the range of A^*. In this case we compute $A^* = QR$ with a (n,m) unitary matrix Q and an upper triangular matrix R. Putting $x = Qz$ we see that x is uniquely determined by

$$R^*z = b \quad , \quad x = Qz \quad . \tag{3.2}$$

Again the Householder process can be employed.

In principle, the method for solving overdetermined systems described above can also be used to solve the regularized normal equations (1.4) by applying it to the overdetermined system

$$\begin{pmatrix} A \\ \gamma I_n \end{pmatrix} x = \begin{pmatrix} b \\ 0 \end{pmatrix}$$

consisting of $m + n$ equations. However, considerable savings can be obtained by a method due to Elden (77). First we remark that minimizing (1.3) is equivalent to minimizing

$$||U A V^*y - U b||^2 + \gamma^2 ||y||^2$$

where U, V are unitary matrices and $x = V^*y$. We show that by a suitable choice of U, V, the matrix $B = UAV^*$ can be made bidiagonal, i.e. only the elements in positions (i, i), $(i, i+1)$ are possibly different from 0. U, V are again constructed by a Householder process. To fix ideas, let $m \geq n$. We put

$$U = U_n \ldots U_1, \quad V^* = V_1 \ldots V_{n-1}, \quad A_k = U_k \ldots U_1 A V_1 \ldots V_k$$

where

$$U_k = \left[\begin{array}{c|c} I_{k-1} & \\ \hline & I_{m-k+1} - 2u_k u_k^* \end{array}\right], \quad V_k = \left[\begin{array}{c|c} I_k & \\ \hline & I_{n-k} - 2v_k v_k^* \end{array}\right]$$

We determine the vectors u_k, v_k such that

$$A_k = \left[\begin{array}{c|c} B_k & \\ \hline & A_k' \end{array}\right] \tag{3.3}$$

where B_k is a bidiagonal $(k+1, k+1)$ - matrix. This can be achieved in the following way: First choose u_1 such that the first column of $U_1 A$ is zero below the diagonal. This step is identical to the first step in the QR - decomposition described above. Then choose v_1 such that the first row of $A_1 = U_1 A V_1$ is zero in columns 3 to n. This step is quite analogous to a Householder step. Since, due to the special shape of V_1, the zeros in the first column of $U_1 A$ are retained, A_1 has the form (3.3) for $k = 1$. Now we repeat the whole process, eliminating the second column below the (2,2) - element by a left multiplication with U_2, followed by eliminating the elements 4 to n in row 2 by a right multiplication with V_2. After $n-1$ steps we arrive at a matrix A_{n-1} which is bidiagonal in rows 1 to n. Eliminating the elements in column n below the diagonal by a left multiplication with U_n finishes the bidiagonalization of A. Since we carried out essentially a double

Householder process, the operation count for the bidiagonalization is $\frac{4}{3}n^3 + O(n^2)$ for $n = m$.

Now we are left with solving the overdetermined system

$$\begin{pmatrix} B \\ \gamma I_n \end{pmatrix} y = \begin{pmatrix} Ub \\ 0 \end{pmatrix}, \quad B \text{ bidiagonal } (m,n)\text{-matrix}, \qquad (3.4)$$

in the generalized sense. Due to the sparsity pattern of the matrix of (3.4) we can use Givens rotations to reduce (3.4) to upper triangular form. A Givens rotation $R(i,j)$ is a unitary matrix, deviating from the unit matrix only in four elements $r_{ii} = r_{jj} = c$, $r_{ji} = -r_{ij} = s$ where $c^2 + s^2 = 1$. Applying $R(i,j)$ to a vector changes only components i, j of that vector, and c, s can be determined so as to anihilate either of these components. (3.4) is left multiplied by $R(1,m+1)$ (anihilates $(m+1,1)$), $R(m+1,m+2)$ (anihilates $(m+1,2)$), $R(2,m+2)$ (anihilates $(m+2,2)$), $R(m+2,m+3)$ (anihilates $(m+2,3)$) and so on. For $n = m = 4$ the triangulation process can be visualized as follows:

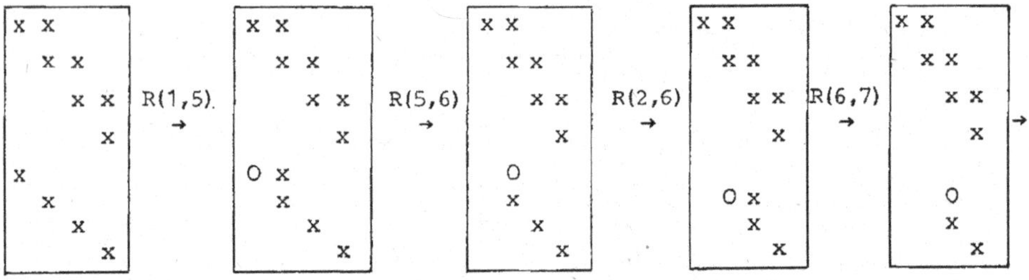

In this scheme, freshly introduced zeros are shown explicitly, while x denotes a generic non zero element. Due to the special structure of the matrices, each Givens rotation can be done in $O(1)$ operations. The work estimate for the triangulation process is therefore $O(n)$. Since the resulting triangular system is bidiagonal, the same applies to the computation of y, while $x = V^*y = V_1 \dots V_{n-1}y$ can be done in $n^2 + O(n)$ operations. Hence the total solution process for the regularized normal equations requires $\frac{4}{3}n^3 + O(n^2)$ operations, plus $n^2 + O(n)$ operations for each value of γ.

4. COMPUTING THE SVD

In the following we describe essential features of the Golub - Reinsch (70) SVD algorithm which is implemented as subroutine LSVDF in the IMSL library and as FO2WCF in the NAG library.

Let A be a (m,n) - matrix with $m \geq n$. It has been shown in section 3 how to compute unitary matrices U, V such that $B = UAV^*$ is bidiagonal. Obviously, B has the same singular values as A . Therefore we consider only bidiagonal matrices B .

The singular values of B are the eigenvalues of the symmetric tridiagonal matrix $M = B^*B$. There are very efficient methods for computing the eigenvalues of such a matrix, most notably the QR - algorithm of Francis (61). Ignoring the shifts, which are necessary to guarantee fast convergence, it reads

$$\begin{aligned} M_1 &= M \\ M_k &= QR \quad , \quad Q \text{ unitary, } R \text{ upper triangular,} \\ M_{k+1} &= RQ \quad . \end{aligned} \tag{4.1}$$

Subject to certain conditions, the diagonal elements of M_k converge to the eigenvalues of M .

Unfortunately, computing $M = B^*B$ squares the condition number, and we get the same sort of problems as in forming A^*A. However, there is a very nice trick which allows to do the iteration (4.1) without computing B^*B.

It follows from (4.1) that

$$M_{k+1} = Q^*Q\,R\,Q = Q^*\,M_k\,Q \quad , \tag{4.2}$$

i.e. M_{k+1} and M_k are similar. Now consider the iteration

$$\begin{aligned} B_1 &= B \\ B_{k+1} &= S^*\,B_k\,T \,, \quad S, T \text{ unitary.} \end{aligned} \tag{4.3}$$

It follows that

$$M_{k+1} = B_{k+1}^{*} B_{k+1} = T^{*} B_{k}^{*} S S^{*} B_k T = T^{*} M_k T \quad . \tag{4.4}$$

Now we make use of the following lemma of Francis (61).

<u>LEMMA 4.1:</u> Let A be a square matrix, let M be a tridiagonal matrix whose elements below the diagonal do not vanish, and let Q be unitary. If

$$M = Q^{*} A Q \quad ,$$

then Q is uniquely determined, up to a diagonal matrix with entries ± 1, by A and the first column of Q.

<u>PROOF:</u> Let $M = (m_{ij})$, $Q = (q_1,\ldots,q_n)$.
The first column of $QM = AQ$ reads

$$m_{11} q_1 + m_{21} q_2 = A q_1 \quad .$$

It follows that

$$m_{11} = q_1^{*} A q_1 \quad , \quad m_{21} = \pm \| A q_1 - m_{11} q_1 \| \quad ,$$

hence $\pm q_2$ is determined by A and q_1. The same reasoning is applied recursively for columns 3 to n. □

The idea is now to choose S, T in (4.3) such that

(i) B_{k+1} is bidiagonal (as is B_k),

(ii) the first column of T coincides, up to a factor of ± 1, with the first column of Q from (4.1). (4.5)

Then, by applying the lemma to (4.2), (4.4), the iterations (4.1), (4.3) are equivalent, provided the elements below the diagonal of M_{k+1} do not vanish.

Matrices S, T defining the transition $B_k \to B_{k+1} = S^{*} B_k T$ and satisfying (4.5) are constructed in the following way. Put

$$S = S_2 \ldots S_n \quad , \quad T = T_2 \ldots T_n$$

where each S_i, T_i is a Givens rotation in the plane $(i-1,i)$. Since all the matrices M_k are tridiagonal (preserving the tridiagonal shape of a matrix is a general property of the QR-algorithm), the first column of Q has nontrivial elements only in the first two rows. Thus we can choose T_2 such that it has, up to a factor of ± 1, the same first column as Q. Hence condition (ii) of (4.5) is satisfied. Condition (i) is achieved by the following "chasing" process:

Possibly $(B_kT_2)_{21} \neq 0$. Choose S_2 s.t. $(S_2B_kT_2)_{21} = 0$.

Possibly $(S_2B_kT_2)_{13} \neq 0$. Choose T_3 s.t. $(S_2B_kT_2T_3)_{13} = 0$.

Possibly $(S_2B_kT_2T_3)_{32} \neq 0$. Choose S_3 s.t. $(S_3S_2B_kT_2T_3)_{32} = 0$.

⋮

We visualize the chasing process for $n = 5$:

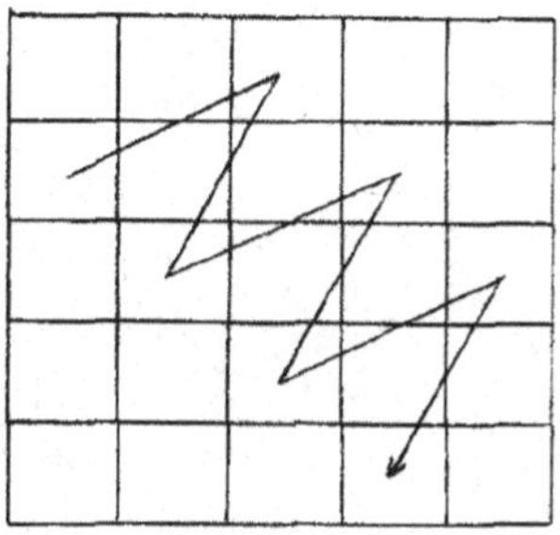

As far as numerical stability is concerned, the replacement of (4.1) by (4.3) is the crucial point of the Colub - Reinsch algorithm. We do not discuss details, such as introducing shifts, taking care of vanishing elements in B_k, and stopping criteria. We refer the reader to the original paper of Golub - Reinsch (70) and to the detailed discussion in Lawson - Hanson (74).

5. ITERATIVE METHODS

There is a variety of iterative method for computing generalized, regularized, maximum entropy, constrained solution to $Ax = b$ with a (m,n) - matrix A. We give only a survey concentrating on the behaviour of these methods for ill-posed problems.

5.1 The Landweber (51) iteration

Here, a sequence x^t, $t=0,1,\ldots$ is computed from

$$x^{t+1} = x^t + \gamma^2 A^*(b - Ax^t) \quad , \tag{5.1}$$

$\gamma > 0$ being a parameter. We analyse (5.1) with the help of the SVD (1.6) of A. Let $x^o = 0$. Then,

$$x^t = \sum_{k=1}^{p} c_k^t v_k \quad ,$$

where the c_k^t satisfy the recursion

$$c_k^{t+1} = (1 - \gamma^2\sigma_k^2)c_k^t + \gamma^2\sigma_k(b,u_k) \quad .$$

From $c_k^o = 0$ we get

$$c_k^t = F_t(\sigma_k)\sigma_k^{-1}(b,u_k) \quad ,$$

$$F_t(\sigma) = 1 - (1-\gamma^2\sigma^2)^t \quad .$$

Hence

$$x^t = \sum_{k=1}^{p} F_t(\sigma_k)\sigma_k^{-1}(b,u_k)v_k \quad .$$

We see that Landweber's iteration can be viewed as a special case of digital filtering. It converges to A^+b provided that $\gamma\sigma_1 < 1$. We also note that the speed of convergence of the contribution of v_k depends on the size of σ_k. For σ_k large, the convergence is fast, but for small σ_k's, the convergence is slow. Thus in the early iterates, the contribution of the large singular values are well represented, while the contribution of the small singular values appear only later in the iteration process. This shows that stopping the iteration after a finite number of step has the same effect as regularization. Carrying out too many steps may destroy the accuracy already obtained since for t large, the iterates pick up the contributions of the small singular values which are very sensitive with respect to data errors. This semi-convergence of iterative methods for ill-posed problems in the presence of noise is very typical.

For extensions of the Landweber iteration see Strand (74).

5.2 The conjugate gradient method (CG)

The Landweber iteration computes first those parts of the solution which belong to the large singular values. The CG algorithm has a similar but even more favourable property.

CG is a method for minimizing $||Ax-b||$. It works as follows. Choose x^0 and put $d^0 = b - Qx^0$, $Q = A^*A$. Then compute x^t, d^t recursively from

$$x^{t+1} = x^t + \alpha^t d^t \quad , \quad \alpha^t = - \frac{(g^t, d^t)}{(d^t, Qd^t)} \quad , \quad g^t = -A^*b + Qx^t$$

and

$$d^{t+1} = - g^{t+1} + \beta^t d^t, \quad \beta^t = \frac{(g^{t+1}, Qd^t)}{(d^t, Qd^t)} \quad .$$

Note that CG forms explicitly A^*A. Therefore it should not be used if singular values of size $10^{-t/2}$, t the number of decimals of the computer, are relevant.

The CG algorithm has the following nice optimality property, see Luenberger (73), chapt. 8.4.

<u>LEMMA 5.1:</u> Let $x^0 = 0$. Then,

$$x^t = P_t(Q)A^*b$$

with a polynomial P_t, of degree $< t$. Among all polynomials P of degree $< t$, P_t minimizes

$$(x, Q(I - QP(Q))^2 x)$$

where $x = A^+b$.

In terms of the SVD, the lemma says that

$$x^t = P_t(Q)A^*b = \sum_{k=1}^{p} \sigma_k P_t(\sigma_k^2)(b, u_k)v_k \quad , \tag{5.2}$$

and P_t minimizes

$$\sum_{k=1}^{p} (x,v_k)^2 \sigma_k^2 (1 - \sigma_k^2 P(\sigma_k^2))^2$$

among all polynomials of degree $< t$. This means that $1 - \sigma_k^2 P_t(\sigma_k^2)$ is small for $(x,v_k)^2 \sigma_k^2$ large, i.e. $\sigma_k P_t(\sigma_k^2)$ is close to σ_k^{-1} in some sense. Comparing (5.2) with

$$x = A^+ b = \sum_{k=1}^{p} \sigma_k^{-1} (b,u_k) v_k$$

we see that CG provides a good approximation in those subspaces $sp\{v_k\}$ for which $(x,v_k)^2 \sigma_k^2$ is large. This means that CG not only tends to pick up soon the contributions of the large singular values, it also takes into account the size of this contribution as measured by $(x,v_k)^2$. For instance, it does not work on the component v_k if v_k is not really contained in x, irrespective of the size of σ_k. This has been first observed by Johnson (79).

5.3 Algebraic reconstruction technique (ART)

For ART we write $Ax = b$ as

$$(a_j, x) = b_j \quad , \quad j = 1, \ldots, m \quad , \quad \|a_j\| = 1 \quad ,$$

where $a_j \in \mathbb{R}^n$ are the rows of A. With P_j the orthogonal projection on the j-th of these hyperplanes, i.e.

$$P_j x = x + (b_j - (a_j, x)) a_j$$

we put, with $0 < \omega < 2$ a relaxation parameter,

$$P_j^{\omega} = (1 - \omega) I + \omega P_j \quad , \quad P^{\omega} = P_m^{\omega} \cdots P_1^{\omega} \quad .$$

Then, the ART iterates are $x^{t+1} = P^{\omega} x^t$. For $\omega = 1$, we obtain the classical Kaczmarz (37) method which simply projects orthogonally on the hyperplanes defining $Ax = b$.

There is a vast literature on ART, in particular related to medical imaging, see Herman (80), chapt. 11. The convergence analysis can be done within the frame work of the SOR theory of numerical analysis, since ART is essentially the same as the familiar SOR method for the system $AA^* u = g$, as has been pointed out by Björck - Elfving (79).

<u>THEOREM 5.2:</u> Let $0 < \omega < 2$. Then, for $x^0 \in R(A^*)$ (e.g. $x^0 = 0$), the ART iterates converge to the unique solution $x_\omega \in R(A^*)$ of

$$A^*(D + \omega L)^{-1}(b - Ax_\omega) = 0$$

where D is the diagonal and L the lower left part of AA^*. If $Ax = b$ has a (classical) solution, then x_ω is the solution with minimal norm. Otherwise,

$$x_\omega = A^+b + O(\omega) \quad .$$

This theorem is essentially due to Censor et al. (83).

Even though the theorem looks very much like the usual SOR theorems of numerical analysis, there are some striking differences. The parameter ω, as well as the ordering of the equations in $Ax = b$, not only influences the speed of convergence, but also (in the inconsistent case) the limit. In the next section we shall study a special case more closely.

Iterative methods can also be used to solve the constrained quadratic programming problem (see Cryer (71)) and the maximum entropy problem (see Herman - Lent (76)).

6. APPLICATIONS TO COMPUTERIZED TOMOGRAPHY (CT)

In CT one has to compute a function f in $\mathbb{R}^2$ from a finite set of its one-dimensional projections, see Herman (80), Natterer (86). We assume f to be supported in the unit disk. The projections onto the directions $\theta_1, \ldots, \theta_p$ are

$$g_j = A_j f(s) = \int f(s\theta_j + t\,\theta_j^\perp)\,dt \quad . \tag{6.1}$$

One of the standard methods of computing f from $g_1, \ldots, g_p$ is ART. The convergence of - suitably discretized versions of - ART has been shown in section 5. However, a general convergence theorem such as Theorem 5.2 does not mean much in a specific case. Fortunately, a much deeper convergence analysis is possible in CT.

To begin with we consider the continuous analogue of ART for (6.1):

Let P_j be the orthogonal projection in $L_2(|x| < 1)$ onto the subspace $g_j = A_j f$, and let

$$P_j^\omega = (1-\omega)I + \omega P_j \quad , \quad P^\omega = P_p^\omega \cdots P_1^\omega \quad .$$

Then, ART reads

$$f^{t+1} = P^\omega f^t \quad , \quad f^0 = 0 \quad .$$

Assuming (6.1) to be consistent, f^t converges to the minimum norm solution f^+ of (6.1), provided that $0 < \omega < 2$.

In order to get more information about the speed of convergence we put $e^t = f^+ - f^t$. Then,

$$e^{t+1} = Q^\omega e^t \quad , \tag{6.2}$$

where

$$Q_j^\omega = (1-\omega)I + \omega Q_j \quad , \quad Q^\omega = Q_p^\omega \cdots Q_1^\omega$$

and Q_j is the orthogonal projection onto the subspace $A_j f = 0$.

Let U_m denote the Chebysheff polynomial of degree m and let $U_{m,j}(x) = U_m((x \cdot \theta_j))$. Let

$$\mathcal{U}_m = \mathrm{sp}\,\{U_{m,1}, \ldots, U_{m,p}\} \quad .$$

Hamaker - Solmon (78) have obtained the following theorem.

<u>LEMMA 6.1:</u> $\mathcal{U}_m$ is an invariant subspace of Q^ω.

The lemma permits to study (6.2) on each $\mathcal{U}_m$ separately. It is a matter of routine to compute the norm $\rho_m(\omega)$ of Q^ω on $\mathcal{U}_m$. One can show that the parts of f which are in $\mathcal{U}_m$ for $m \geq p$ are not determined by (6.2), i.e. there is no point in studying the speed of convergence of (6.2) in $\mathcal{U}_m$ for $m \geq p$, see Natterer (86), chapt. V.4.

We did the following numerical experiments:

(a) We computed $\rho_m(\omega)$ for $p = 32$ directions $\theta_j = \begin{pmatrix} \cos\varphi_j \\ \sin\varphi_j \end{pmatrix}$,

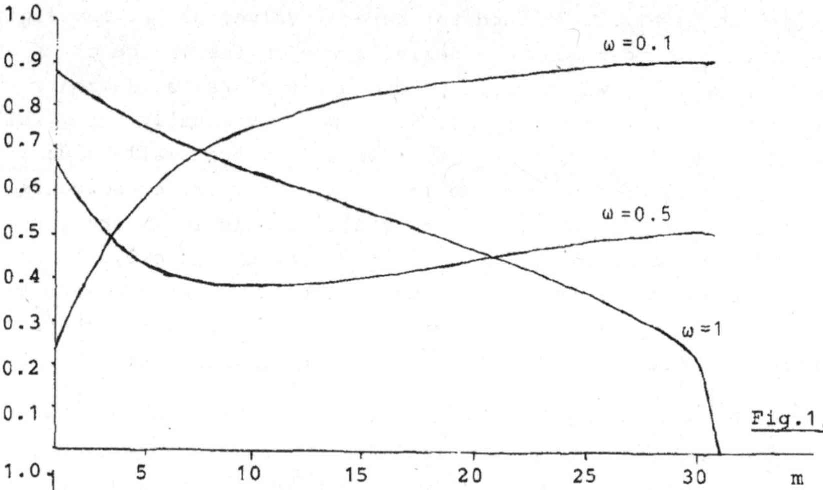

Fig.1.

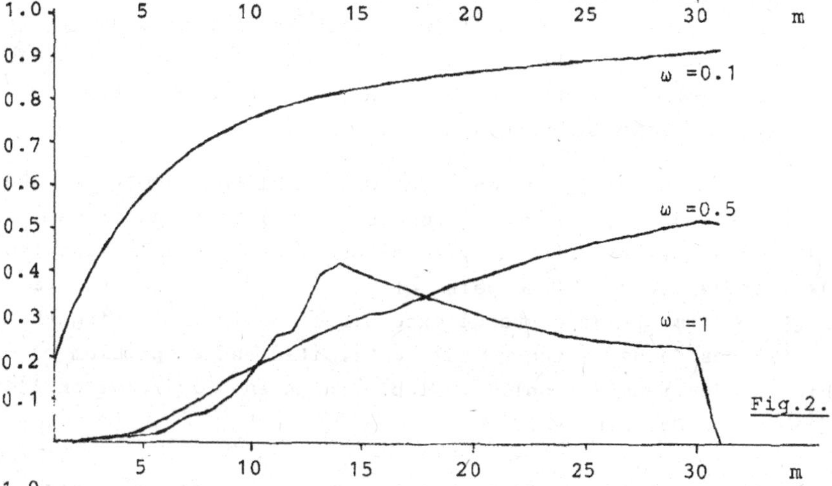

Fig.2.

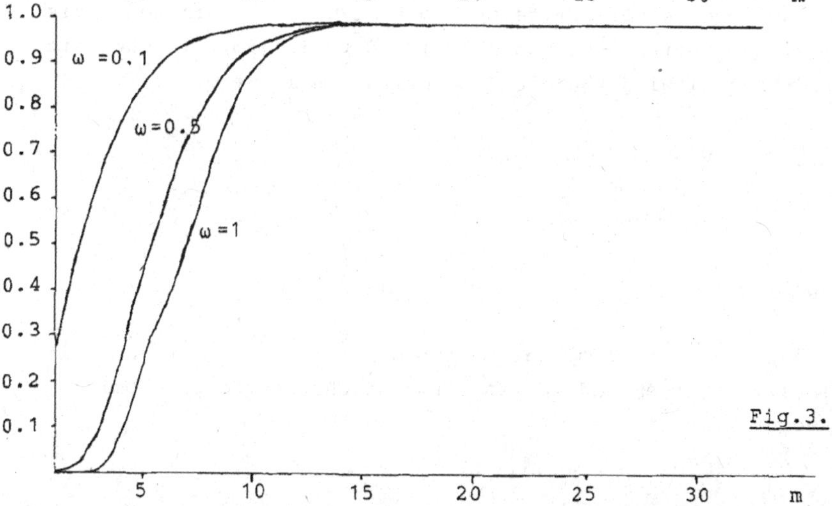

Fig.3.

$\varphi_j = \pi(j-1)/p$, $j=1,\dots,32$, and for several values of ω, see fig. 1. We see that $\rho_m(\omega)$ depends in a decisive way on the choice of ω. For ω large (e.g. $\omega=1$, which corresponds to the classical Kaczmarz method) ρ_m decreases from $\rho_o = 0.86$ to ρ_{p-1} virtually zero, while for ω small (e.g. $\omega = 0.1$), ρ_m increases from the small value $\rho_o = 0.26$ to $\rho_{p-1} = 0.90$. This has the following practical consequences for the iterates f^t: For ω large, the small details in f are picked up quickly, while the smooth parts of f converge only slowly. For ω small, the opposite is the case. Thus, choosing ω has the same effect as regularization on the early iterates. This accounts for the surprisingly low values of ω which are being used in practice.

(b) Again we chose 32 equally spaced directions, but we permuted them in a random way, see fig. 2. Now the $\rho_m(\omega)$ are much more favourable for ω large, with values virtually zero on the first few $\mathcal{U}_m$, with maximum significantly smaller than for the natural order in fig. 1. In fact it is observed in practice that ART performs much better for orderings other than the natural one.

(c) Again we chose 32 directions, but we restricted them to an angular range of 150°, and we permuted them (as in (b)) in a random way, see fig. 3. Now ρ_m is significantly less than 1 only on the first few of the $\mathcal{U}_m$, irrespective of the value of ω. This means that the parts of f in $\mathcal{U}_m$ for m large converge exceedingly slow, indicating the severe ill-posedness of the so called "limited angle" problem of CT, in which projections are only available in an angular range smaller than 180°, see Davison (83).

There have been several attempts to deal with the limited angle problem, see Davison - Grünbaum (81) and Louis (80). In the latter paper, the following range characterization is used.

<u>LEMMA 6.2:</u> For $s = 0,1,\dots,$

$$\int s^m g_j(s)ds$$

is a polynomial of degree $\le m$ in θ_j.

This can be used to compute g_j for directions θ_j for which it is not available. First expand g_j in terms of Chebysheff polynomials U_m of the second kind, i.e.

$$g_j(s) = (1-s^2)^{1/2} \sum_{m=0}^{\infty} U_m(s) g_{mj} \quad ,$$

$$g_{mj} = \frac{2}{\pi} \int_{-1}^{+1} g_j(s) U_m(s)\,ds \quad .$$

Since, by the lemma, g_{mj} is a polynomial of degree m in θ_j, the missing directions can be computed simply by extrapolation whenever the number p of given directions is sufficiently large to determine the coefficients of g_{mj}, i.e. $p > m$. The extrapolation is done by solving an overdetermined severely ill-conditioned linear system. Tikhonov - Phillips regularization and SVD have successfully been used in cases where the missing angular range is not too large, i.e. less than 30^0.

REFERENCES

Ben-Israel, A. - Greville, T.V.E.: Generalized Inverses: Theory and Applications. Wiley Interscience 1974

Bertero, M.: Problemi lineari non ben posti e metodi di regolarizzazione. Publicazioni dell'istituto di analisi globale e applicazioni, Florence 1982

Björck, Å.: Solving Linear Least Squares Problems by Gram-Schmidt Orthogonalization, BIT 7, 1-21 (1967)

Björck, Å. - Elden, L.: Methods in Numerical Algebra for Ill-Posed Problems, Technical Report LiTH-MAT-R33-1979, LINKÖPING UNIVERSITY 1979

Björck, Å. - Elfving, T.: Accelerated Projection Methods for Computing Pseudoinverse Solutions of Linear Equations, BIT 19, 145-169 (1979)

Censor, Y. - Eggermont, P.P.B. - Gordon, D.: Strong Underrelaxation in Kaczmarz's Method for Inconsistent Systems, Numer. Math. 41, 83-92 (1983)

Cryer, C.W.: The Solution of a Quadratic Programming Problem Using Systematic Overrelaxation, SIAM J. Control 9, 385-392 (1971)

Davison, M.E.: The Ill-Conditioned Nature of the Limited Angle Tomography Problem, SIAM J. Appl. Math. 43, 428-448 (1983)

Davison, M.E. - Grünbaum, F.A.: Tomographic Reconstructions with Arbitrary Directions. Comm. Pure Applied Math. 34, 77-120 (1981)

Eldén, L.: A Note on the Computation of the Generalized Cross - Validation Function for Ill-Conditioned Least Squares Problems, BIT 24, 467-472 (1984)

Eldén, L.: Algorithms for the Regularization of Ill-Conditioned Least Squares Problems, BIT 17, 134-145 (1977)

Engl, H.W.: Regularization and Least Squares Collocation, in: Hämmerlin, G. - Hoffmann, K.-H.: Improperly Posed Problems and their Numerical Treatment, ISNM, Birkhäuser 1983

Francis, J.G.F.: The QR Transformation: A Unitary Analogue of the LR Transformation, Parts I and II. Comp. J. 4, 265-272 and 332 - 345 (1961)

Golub, G.H. - Saunders, M.A.: Linear Least Squares and Quadratic Programming, in: Abadie, J. (ed.): Integer and Nonlinear Programming. North-Holland 1970

Golub, G.H. - Reinsch, C.: Singular Value Decomposition and Least Squares Solutions, Numer. Math. 14, 403-420 (1970)

Golub, G.H.: Numerical Methods for Solving Least Squares Problems, Numer. Math. 7, 206-216 (1965)

Groetsch, C.W.: The Theory of Tikhonov Regularization for Fredholm Equations of the First Kind. Pitman 1984

Hamaker, C. - Solmon, D.C.: The Angles between the Null-Spaces of X-Rays, J. Math. Anal. Appl. 62, 1-23 (1978)

Herman, G.T.: Image Reconstruction from Projections. Academic Press 1980

Herman, G.T. - Lent, A.: Iterative Reconstruction Algorithms, Comput. Biol. Med. 6, 273-294 (1976)

Isaacson, E. - Keller, H.B.: Analysis of Numerical Methods. Wiley 1966

Ivanov, V.V.: The Theory of Approximate Methods and their Application to the Numerical Solution of Singular Integral Equations. Nordhoff International Publishing 1976

Johnsson, C.: On Finite Element Methods for Optimal Control Problems, Part II, Ill-Posed Problems, Report 79-04R, Dept. Computer Sciences, University of Gothenburg

Kaczmarz, S.: Angenäherte Auflösung von Systemen linearer Gleichungen, Bull. Int. Acad. Pol. Sci. Lett. A., 355-357 (1937)

Landweber, L.: An Iterative Formula for Fredholm Integral Equations of the First Kind, Amer. J. Math. 73, 615-624 (1951)

Lavrentiev, M.M. - Romanov, V.G. - Sisatskij, S.P.: Problemi non ben posti in Fisica Matematica e Analisi. Publicazioni dell'istituto di analisi globale e applicazioni, Florence 1983

Lawson, C.L. - Hanson, R.J.: Solving Least Squares Problems. Prentice-Hall 1974

Louis, A.K.: Picture Reconstruction from Projections in Restricted Range, Math. Meth. in the Applied Sci. 2, 209-220 (1980)

Luenberger, D.G.: Introduction to Linear and Nonlinear Programming. Addison-Wesley 1973

Morozov, V.A.: Methods for Solving Incorrectly Posed Problems. Springer 1984

Nashed, M.Z. (ed.): Generalized Inverses and Applications. Academic Press 1976

Nashed, M.Z.: On Moment-Discretization and Least-Squares Solutions of Linear Integral Equations of the First Kind, J. Math. Anal. Appl. 53, 359-366 (1976)

Natterer, F.: Regularisierung schlecht gestellter Probleme durch Projektionsverfahren, Numer. Math. 28, 329-341 (1977)

Natterer, F.: Discretizing Ill-Posed Problems, Publicazioni dell' istituto di analisi globale e applicazioni, Firenze 1983

Natterer, F.: The Mathematics of Computerized Tomography. Wiley - Teubner 1986

Smith, C.R. - Grandy, W.T.: Maximum Entropy and Bayesian Methods in Inverse Problems. Reidel 1985

Strand, O.N.: Theory and Methods Related to the Singular Value Expansion and Landweber's Iteration for Integral Equations of the First Kind, SIAM J. Numer. Anal. 11, 798-825 (1974)

Tikhonov, A.N. - Arsenin, V.Y.: Solution of Ill-Posed Problems. Winston & Sons 1977

Varah, J.M.: A Practical Examination of Some Numerical Methods for Linear Discrete Ill-Posed Problems, SIAM Review 21, 100-111 (1979)

Wedin, P.Å.: Perturbation Theory for Pseudo-Inverses, BIT 13, 217-232 (1973)

Wilkinson, J.H.: Rounding Errors in Algebraic Processes. Prentice-Hall 1974

Wilkinson, J.H.: The Algebraic Eigenvalue Problem. Clarendon Press 1965

REGULARIZATION WITH LINEAR EQUALITY CONSTRAINTS*

C. W. Groetsch
University of Cincinnati

ABSTRACT

This paper is an exposition of the basic theory of regularization for an ill-posed linear operator equation subject to a linear operator constraint. The ordinary theory of regularization is subsumed as a special case. The theory is developed by changing the geometric structure of the underlying Hilbert space and invoking well known results on generalized inverses. In addition to the basic theory, the convergence of an approximation method, including a finite element implementation, is considered.

1. INTRODUCTION.

The theory of regularization for an ill-posed linear operator equation of the first kind

$$Kx = g \qquad (1)$$

is well developed (see [6] for a personal perspective of this theory). The prototype of such an equation is the Fredholm integral equation of the first kind

$$\int_a^b k(s,t)x(t)dt = g(s) \ ,$$

where $k(\cdot,\cdot)$ is a given square integrable kernel and g is a given square integrable function. Such equations occur frequently in the study of inverse problems and the major difficulty in their practical solution derives from the fact that the dependence of the solution x on the data g is generally discontinuous. The function g is usually the result of measurements and hence is only imprecisely known. Noise in the data is then amplified by the solution process and manifests

*Partially supported by a grant from the National Science Foundation.

itself as large instabilities in the computed solution. The method of Tikhonov regularization is a device for damping these instabilities in order to obtain useful approximations to the solution.

It sometimes happens that in addition to satisfying (1), it is desirable that the solution x satisfy an additional condition of the form

$$Lx = f \tag{2}$$

where L is a given linear operator. For example, if the function x which satisfies the integral equation above is also known to be a density function, then in addition it should satisfy

$$\int_a^b x(t)dt = 1$$

(see [7] for a medical example involving the modeling of the human liver). Other examples of problems which require that the function x satisfy operator equations of the type (1) - (2) come to mind. For example, in [1, Section 6.6] the earth's density distribution is given as the solution of a pair of integral equations involving the mean density and moment of inertia of the earth.

In general, the simultaneous satisfaction of (1) and (2) may be prohibitively restrictive. Indeed, it is generally the case that (1) has no solution in the usual sense and a least squares solution must be sought. We will seek, among all least squares solutions of (1), one which also satisfies (2) in the least squares sense.

A well known example of a problem of this type is that of data fitting with a smooth spline. For appropriate function spaces, one wishes to solve

$$Kx = g$$

where $Kx = [x(t_1),\ldots,x(t_n)]^T$ for given nodes $\{t_i\}_{i=1}^n$ and g is a given vector in R^n, while also requiring that x be a least squares solution of $Lx = 0$ where L is a linear differential operator, e.g., $Lx = x''$.

Minamide and Nakamura [9] have studied a problem in control theory that falls within the context of this article. Consider the linear dynamical system

$$\begin{aligned} x'(t) &= A(t)x(t) + B(t)u(t) \ , \quad t_0 < t < t_1 \\ x(t_0) &= x_0 \end{aligned} \tag{3}$$

where $x(t) \in R^n$ is a state vector, $u(t) \in R^m$ is a control vector and $A(t)$, $B(t)$ are matrices of appropriate dimensions. Let U be the Cartesian product of m copies of $L^2[t_0,t_1]$ with inner product

$$(u,v) = \int_{t_0}^{t_1} \langle u(t),R(t)v(t)\rangle dt$$

where $\langle\cdot,\cdot\rangle$ is the Euclidean inner product and $R(t)$ is a given mxm positive definite matrix. Let X be the Cartesian product of n copies of $L^2[t_0,t_1]$ with inner product

$$(x,y) = \int_{t_0}^{t_1} \langle x(t),y(t)\rangle dt \ .$$

Suppose $x_d \in X$ is given and consider the cost functionals

$$J_1(u) = \|x(t_1)-x_d(t_1)\|^2$$

and

$$J_2(u) = \|u\|^2 + (Q(x-x_d),Q(x-x_d))$$

where $Q(t)$ is a given matrix function. Minamide and Nakamura consider the optimization problem of finding

$$\hat{u} \in B = \{v \in U : J_1(v) \leq J_1(u), \text{ all } u \in U\}$$

such that $J_2(\hat{u}) \leq J_2(v)$ for all $v \in B$.

If $\Phi(t,s)$ is the transition matrix for the system (3), then

$$x(t) = \Phi(t,t_0)x_0 + \int_{t_0}^{t} \Phi(t,s)B(s)u(s)\,ds .$$

Define $T:U \to X$ by

$$(Tu)(t) = Q(t) \int_{t_0}^{t} \Phi(t,s)B(s)u(s)ds$$

and $K:U \to R^n$ by

$$Ku = \int_{t_0}^{t_1} \Phi(t_1,s)B(s)u(s)ds .$$

Let $\phi(t) = Q(t)(x_d(t)-\Phi(t,t_0)x_0)$ and $g = x_d(t_1)-\Phi(t_1,t_0)x_0$.
Then $J_1(u) = \|Ku-g\|^2$ and

$$J_2(u) = \|u\|^2 + \|Tu-\phi\|^2 .$$

Finally, let L be the product operator

$$(I,T):U \to U \times X$$

where I is the identity operator, and let $f = \{0,\phi\}$, then the required $\hat{u}$ is the least squares solution of $Ku = g$ which also satisfies $Lu = f$ in the least squares sense.

In finite dimensional spaces least squares problems with linear equality constraints have been studied by Eldén [4] and others. Morozov [10] has considered the problem in a general Hilbert space setting. The presentation given here is based on [3] (N.B. in [3] the notation for K and L is the reverse of that given here).

2. BASIC THEORY

For simplicity we consider first the case of bounded operators. Suppose $K:H_1 \to H_2$ and $L:H_1 \to H_3$ are bounded linear operators on Hilbert spaces. For given $g \in H_2$ and $f \in H_3$, we wish to find among all least squares solutions of (1) a least squares solution of (2). Now (1) has a least squares solution if and only if $g \in \mathcal{D}(K^\dagger) = R(K) + R(K)^\perp$, where $K^\dagger$ is the Moore-Penrose generalized inverse of K. The set of all least squares solutions of (1) is then $K^\dagger g + N(K)$ where $N(K)$ is the nullspace of K (see e.g. [5]). We therefore wish to find $\hat{x} \in H_1$ such that

$$\|L\hat{x}-f\| = \inf\{\|Lu-f\| : u \in K^\dagger g + N(K)\} .$$

We shall assume that $N(K) \cap N(L) = \{0\}$ and that the product operator $(K,L):H_1 \to H_2 \times H_3$ has closed range. (Note that $\hat{x} = K^\dagger g$ in the special case when $L = I$ and $f = 0$). The vector $\hat{x}$ will be called the restricted pseudosolution of (1) - (2). Because of the assumption on the nullspaces, at most one such vector $\hat{x}$ exists.

Let $\bar{L}$ designate the operator L restricted to N(K). It is easy to see that, under the assumptions above, the range of $\bar{L}$, $R(\bar{L})$, is closed and hence $\mathcal{D}(\bar{L}^\dagger) = H_3$. The conditions defining $\hat{x}$ can be written $\hat{x} = K^\dagger g+z$, where $z \in N(K)$ and

$$\|\bar{L}z-(f-LK^\dagger g)\| = \inf\{\|\bar{L}w-(f-LK^\dagger g)\| : w \in N(K)\} .$$

Therefore $z = \bar{L}^\dagger(f-LK^\dagger g)$ and hence

$$\begin{aligned} \hat{x} &= K^\dagger g + \bar{L}^\dagger(f-LK^\dagger g) \\ &= \bar{L}^\dagger f + (I-\bar{L}^\dagger L)K^\dagger g . \end{aligned} \tag{4}$$

Hence for any $f \in H_3$ and any $g \in \mathcal{D}(K^\dagger)$ a restricted pseudosolution exists and is given by (4). A reinterpretation of this decomposition of the restricted pseudosolution $\hat{x}$ will be basic to our development.

Since the product operator (K,L) has closed range, it follows from the open mapping theorem that there is a constant $m > 0$ such that

$$\|Kx\|^2 + \|Lx\|^2 \geq m\|x\|^2 . \tag{5}$$

We can therefore define a new inner product (called the "star" inner product by Locker and Prenter [8]) on H_1 by

$$[x,y] = (Kx,Ky) + (Lx,Ly) .$$

The associated norm will be denoted $|\cdot|$:

$$|x|^2 = [x,x] .$$

We will designate the space H_1 equipped with the inner product $[\cdot,\cdot]$ by $\mathcal{H}_1$ and denote the operator K when operating on $\mathcal{H}_1$ by $\mathcal{K}$.

<u>Proposition 1</u>. $\mathcal{K}^\dagger = (I-\mathcal{L}^\dagger L)K^\dagger$.

Proof. Suppose $g \in \mathcal{D}(\mathcal{K}^\dagger) = \mathcal{D}(K^\dagger)$ and let

$$w = K^\dagger g - \mathcal{L}^\dagger L K^\dagger g .$$

Since $R(\mathcal{L}^\dagger) \subseteq N(K)$, we then have $Kw = KK^\dagger g$ and hence w is a least squares solution of (1). We now show that w has minimal $|\cdot|$-norm, or equivalently that $[w,v] = 0$ for all $v \in N(K)$ and hence $w = \mathcal{K}^\dagger g$.

Now, since $\mathcal{L}\mathcal{L}^\dagger = \overline{Q}$, where $\overline{Q}$ is the orthogonal projector of H_2 onto $R(\mathcal{L})$ (see e.g. [5]), we have for any $v \in N(K)$,

$$[w,v] = (Lw,Lv) = (L(I-\bar{L}^{\dagger}\bar{L})K^{\dagger}g,\ \bar{L}v)$$
$$= ((I-\bar{Q})\bar{L}K^{\dagger}g,\ \bar{L}v) = 0\ .$$

#

Using this result, the decomposition (4) becomes

$$\hat{x} = K^{\dagger}g + \bar{L}^{\dagger}f\ . \tag{6}$$

Our next result shows that the generalized inverse of the restricted operator $\bar{L}$ can be expressed in terms of the generalized inverse of the unrestricted operator L.

Proposition 2. If $f \in D(L^{\dagger})$, then $\bar{L}^{\dagger}f = PL^{\dagger}f$, where P is the orthogonal projector of H_1 onto N(K).

Proof. Suppose $v \in N(K)$, then

$$[\bar{L}^{\dagger}f-L^{\dagger}f,v] = (K(\bar{L}^{\dagger}f-L^{\dagger}f),Kv) + (L(\bar{L}^{\dagger}f-L^{\dagger}f),Lv)$$
$$= (\bar{L}\bar{L}^{\dagger}f,\bar{L}v) - (LL^{\dagger}f,Lv)\ ,$$
$$= (f,\bar{L}v)-(f,Lv) = (f,Lv)-(f,Lv) = 0,$$

since $\bar{L}\bar{L}^{\dagger}$ projects onto $R(\bar{L})$ and $LL^{\dagger}$ projects onto $\overline{R(L)}$.

#

It is now possible to apply the standard approximation theory of generalized inverses to approximate $\hat{x}$. In particular, it is known that

$$(K^*K+\alpha I)^{-1}K^*g \to K^{\dagger}g$$

and

$$(K^*K+\alpha I)^{-1}K^*Kx \to (I-P)x$$

as $\alpha \to 0$, where the convergence is in H_1, (see e.g. [5]).

One method of approximating $\hat{x}$, which is known in the matrix literature as the method of weighting, consists in finding the minimizer x_ε of the functional

$$F(u) = \|Ku-g\|^2 + \varepsilon\|Lu-f\|^2 \quad (\varepsilon>0). \tag{7}$$

A routine calculation shows that

$$\begin{aligned} x_\varepsilon &= (K^*K+\varepsilon L^*L)^{-1}K^*g + (\varepsilon^{-1}K^*K+L^*L)^{-1}L^*f \\ &= \qquad y_\varepsilon \qquad\qquad + \qquad\qquad z_\varepsilon \quad . \end{aligned} \tag{8}$$

Now, it is easy to see that the adjoint of the operator $\mathcal{K}$ is given by

$$\mathcal{K}^* = (L^*L+K^*K)^{-1}K^* .$$

Making this substitution we obtain

$$\begin{aligned} (\mathcal{K}^*\mathcal{K}+\alpha I)^{-1}\mathcal{K}^*g &= (1-\varepsilon)(\varepsilon L^*L+K^*K)^{-1}K^*g \\ &= (1-\varepsilon)y_\varepsilon \end{aligned} \tag{9}$$

where $\varepsilon = \alpha/(1+\alpha)$. It therefore follows that

$$\lim_{\varepsilon\to 0} y_\varepsilon = \lim_{\alpha\to 0}(\mathcal{K}^*\mathcal{K}+\alpha I)^{-1}\mathcal{K}^*g = \mathcal{K}^\dagger g$$

where the convergence is in $\mathcal{H}_1$. From (5) we then have $y_\varepsilon \to \mathcal{K}^\dagger g$ in the norm of H_1 as $\varepsilon \to 0$.

Consider now the vector z_ε. If $f \in \mathcal{D}(L^\dagger)$, then since $L^*LL^\dagger f = L^*f$, we have by (8)

$$\begin{aligned} z_\varepsilon &= (\varepsilon^{-1}K^*K+L^*L)^{-1}L^*LL^\dagger f \\ &= (I-(K^*K+\varepsilon L^*L)^{-1}K^*K)L^\dagger f \\ &= (I-(1+\alpha)(\mathcal{K}^*\mathcal{K}+\alpha I)^{-1}\mathcal{K}^*\mathcal{K})L^\dagger f \end{aligned} \tag{10}$$

where $\varepsilon = \alpha/(1+\alpha)$. Therefore

$$\lim_{\varepsilon\to 0} z_\varepsilon = \lim_{\alpha\to 0} (I-(1+\alpha)(K^*K+\alpha I)^{-1}K^*K)L^\dagger f$$
$$= PL^\dagger f = \overline{L}^\dagger f$$

and hence, by (6), $x_\varepsilon \to \hat{x}$ as $\varepsilon \to 0$ for any $g \in D(K^\dagger)$ and $f \in D(L^\dagger)$.

This result can be extended to any $f \in H_3$ as follows. Since $D(L^\dagger)$ is dense in H_3, any $f \in H_3$ can be approximated to arbitrary accuracy by an $\overline{f} \in D(L^\dagger)$. Then

$$z_\varepsilon - \overline{L}^\dagger f = (\varepsilon^{-1}K^*K+L^*L)^{-1}L^*(f-\overline{f}) +$$
$$\{(\varepsilon^{-1}K^*K+L^*L)^{-1}L^*\overline{f}-\overline{L}^\dagger\overline{f}\} + \overline{L}^\dagger(\overline{f}-f).$$

Now, the operator in the first term above is uniformly bounded for $0 < \varepsilon < 1$, $\overline{L}^\dagger$ is bounded (since $R(\overline{L})$ is closed), and the middle term converges to 0 by the argument above. It then follows that $z_\varepsilon \to \overline{L}^\dagger f$ as $\varepsilon \to 0$ for every $f \in H_3$.

We summarize this discussion as follows:

Proposition 3. Suppose $K:H_1 \to H_2$, $L:H_1 \to H_3$ are bounded linear operators and the product operator (K,L) has closed range and trivial nullspace. Then the problem (1)-(2) has a unique restricted pseudosolution $\hat{x}$ if and only if $g \in D(K^\dagger)$. Moreover, for each $f \in H_3$, $x_\varepsilon \to \hat{x}$ in H_1 as $\varepsilon \to 0$, where x_ε is the minimizer of the functional (7).

The known theory of regularization can also be used to obtain rates of convergence under suitable conditions on the data g and f. For example, it is known (see e.g. [6]) that if $K^\dagger g \in R(K^*K)$, then, by (9), $|(1-\varepsilon)y_\varepsilon - K^\dagger g| = O(\alpha)$ and hence

$$\|y_\varepsilon - K^\dagger g\| = O(\alpha) + O(\varepsilon) = O(\varepsilon) \; .$$

Moreover, this rate is best possible. Also, if

$$K^\dagger K L^\dagger f = (I-P)L^\dagger f \in R(K^*K)$$

then $(K^*K+\alpha I)^{-1}K^*KL^\dagger f$ converges to $(I-P)L^\dagger f$ at a rate $O(\alpha)$ and hence by (10) and Proposition 2, z_ε converges to $\bar{L}^\dagger f$ at a rate $O(\alpha) = O(\varepsilon)$. Therefore under suitable conditions on g and f,

$$\|x_\varepsilon - \hat{x}\| = O(\varepsilon) .$$

3. UNBOUNDED OPERATORS.

The theory developed above can be extended to certain unbounded operators L by using the theory and techniques of Locker and Prenter [8]. In this section we indicate briefly how the extension is carried out. Suppose that $D \subseteq H_1$ is a dense subspace, $L: D \to H_3$ is a linear operator, and $K: H_1 \to H_2$ is a bounded linear operator. We will restrict our attention to the domain D and will also denote the restriction of K to D by K. As before we assume that $N(K) \cap N(L) = \{0\}$ and we now assume that the product operator $(K,L): D \to H_2 \times H_3$ is closed. This is equivalent to assuming that L is closed.

Under these conditions there is an $m > 0$ such that

$$|x|^2 \geq m\|x\|^2 \qquad (11)$$

for all $x \in D$. Indeed, if this is not the case there is a sequence $\{x_n\} \subseteq D$ with $\|x_n\| = 1$ and

$$|x_n|^2 = \|Kx_n\|^2 + \|Lx_n\|^2 \to 0 .$$

Since $\{x_n\}$ is bounded, there is a subsequence $\{x_k\} \subseteq D$ with $x_k \overset{w}{\to} x \in H_1$, $Kx_k \to 0$, and $Lx_k \to 0$. Since (K,L) is closed, its graph is weakly closed and hence $x \in D$, $Kx = 0$ and $Lx = 0$, which contradicts the assumption that $N(K) \cap N(L) = \{0\}$, since $\|x\| = 1$. Using this it is easy to show that D, endowed with the inner product $[\cdot,\cdot]$, is a Hilbert space which we designate by $\mathcal{H}_1$. Moreover, the operators K and L are bounded linear operators on $\mathcal{H}_1$. To distinguish the operator K when acting on the space H_1, we will as before denote it by K. It can now be shown that the restricted pseudosolution $\hat{x}$ exists if and only if $g \in D(K^\dagger)$ and again we have the decomposition (6). As before it follows that

$$\hat{x} = K^\dagger g + PL^\dagger f$$

for $f \in D(L^\dagger)$, where P is the orthogonal projector of H_1 onto $N(K)$.

The argument of Locker and Prenter [8] guarantees that for $1 > \varepsilon > 0$ the operator $(K^*K+\varepsilon L^*L)$ has a uniformly bounded inverse defined on all of H_1 and that for $f \in D(L^*)$ the minimizer $x_\varepsilon = y_\varepsilon + z_\varepsilon$ of all the functional (7) is given by (8). Again the adjoint of K is related to that of K by $K^* = (K^*K+L^*L)^{-1}K^*$. Also, since K is bounded on H_1, we have

$$(K^*K+\beta I)^{-1}K^*g \to K^\dagger g \quad \text{as } \beta \to 0.$$

Moreover, if $f \in D(L^\dagger)$, then the term z_ε has the representation (1) and it follows as before that

$$z_\varepsilon \to PL^\dagger f \quad \text{as} \quad \varepsilon \to 0 .$$

4. DATA PERTURBATIONS AND APPROXIMATIONS.

Consider now the problem (1)-(2) in which only perturbed versions $\tilde{g}$ and $\tilde{f}$ of the data are available. We assume that positive tolerances δ and τ are known such that

$$\|g-\tilde{g}\| \leq \delta \quad \text{and} \quad \|f-\tilde{f}\| \leq \tau \quad .$$

Approximations $\tilde{x}_\varepsilon$ are now formed by minimizing the functional

$$\tilde{F}(u) = \|Ku-\tilde{g}\|^2 + \varepsilon\|Lu-\tilde{f}\|^2 \tag{12}$$

and the challenge is to find conditions so that $\varepsilon \to 0$ and $\tilde{x}_\varepsilon \to \hat{x}$ as $\delta,\tau \to 0$. The corresponding theory for ordinary regularization ($L = I$, $\tilde{f} = 0$) can be found for example in [6]. Some of the basic results given here appear in Morozov [10].

The minimizer x_ε of (7) is characterized by

$$(Kv,Kx_\varepsilon-g) + \varepsilon(Lv,Lx_\varepsilon-f) = 0$$

for all $v \in H_1$. We find it convenient to express this in terms of the inner product

$$[x,y]_\varepsilon = (Kx,Ky) + \varepsilon(Lx,Ly)$$

(with associated norm $|\cdot|_\varepsilon$) by

$$[v,x_\varepsilon-\hat{x}]_\varepsilon = -\varepsilon(Lv,L\hat{x}-f) \tag{13}$$

(see [6, Section 4.2] for the corresponding result for ordinary regularization). If we denote the minimizer of (12) by $\tilde{x}_\varepsilon$, then it follows that

$$[v,\tilde{x}_\varepsilon-\hat{x}]_\varepsilon = -\varepsilon(Lv,L\hat{x}-f)-(Kv,\tilde{g}-g)-\varepsilon(Lv,\tilde{f}-f).$$

Subtracting we have

$$[v,x_\varepsilon-\tilde{x}_\varepsilon]_\varepsilon = (Kv,\tilde{g}-g) + \varepsilon(Lv,\tilde{f}-f) \tag{14}$$

for all $v \in H_1$. Setting $v = x_\varepsilon-\tilde{x}_\varepsilon$, we obtain

$$\begin{aligned}|x_\varepsilon-\tilde{x}_\varepsilon|_\varepsilon^2 &\leq \|K(x_\varepsilon-\tilde{x}_\varepsilon)\|\delta + \|L(x_\varepsilon-\tilde{x}_\varepsilon)\| \varepsilon \tau \\ &\leq \delta^2/2 + \varepsilon\tau^2/2 + |x_\varepsilon-\tilde{x}_\varepsilon|_\varepsilon^2/2 .\end{aligned}$$

Hence

$$\varepsilon|x_\varepsilon-\tilde{x}_\varepsilon|^2 \leq |x_\varepsilon-\tilde{x}_\varepsilon|_\varepsilon^2 \leq \delta^2 + \varepsilon\tau^2$$

and therefore

$$|x_\varepsilon - \tilde{x}_\varepsilon| \leq \delta/\sqrt{\varepsilon} + \tau . \tag{15}$$

Note that this points out that instabilities in $\tilde{x}_\varepsilon$ arise from noise in g, not f. From (15) it follows that $\delta = o(\sqrt{\varepsilon})$ is a sufficient condition for the regularity of the method. Also, under suitable conditions the convergence of x_ε to $\hat{x}$ is $O(\varepsilon)$ and hence for $\varepsilon = C\delta^{2/3}$ an asymptotic convergence rate of $O(\max(\delta^{2/3},\tau))$ is possible for the approximations $\tilde{x}_\varepsilon$.

Finally, we consider briefly a finite element approximation technique for $\hat{x}$. Suppose $V_1 \subseteq V_2 \subseteq \ldots$ is a sequence of subspaces whose union is dense in H_1. Denote by $\tilde{x}_\varepsilon^m$ the minimizer of (12) in V_m. The efficacy of $\tilde{x}_\varepsilon^m$ as an approximation to x depends upon how well the subspaces V_m support the operators K and L, specifically on the convergence of

$$\gamma_m = \|K(I-P_m)\| \quad \text{and} \quad \beta_m = \|L(I-P_m)\|$$

to zero where P_m is orthogonal projector of H_1 onto V_m.

By the same argument as above it follows that $\tilde{x}_\varepsilon^m$ satisfies (14) for all $v \in V_m$. Subtracting these two versions of (14) we obtain

$$[v,\tilde{x}_\varepsilon^m - \tilde{x}_\varepsilon]_\varepsilon = 0 \quad \text{for all } v \in V_m$$

or equivalently $\tilde{x}_\varepsilon^m = \mathcal{P}_m\tilde{x}_\varepsilon$, where $\mathcal{P}_m$ is the $[\cdot,\cdot]_\varepsilon$-orthogonal projector of $\mathcal{H}_1$ onto V_m. It can now be shown as in [6, 4.2.3] that

$$\varepsilon|\tilde{x}_\varepsilon - \tilde{x}_\varepsilon^m|^2 \leq |\tilde{x}_\varepsilon - \tilde{x}_\varepsilon^m|_\varepsilon^2 \leq (\gamma_m^2 + \varepsilon\beta_m^2)\|(I-P_m)\tilde{x}_\varepsilon\|^2 . \tag{16}$$

Suppose that the parameters ε, δ, τ, γ, β all depend on m and converge to zero as $m \to \infty$. Combining (15) and (16) we obtain:

Proposition 4. If $\gamma_m = O(\sqrt{\varepsilon_m})$ and $\delta_m = o(\sqrt{\varepsilon_m})$, then $\tilde{x}^m_{\varepsilon_m} \to \hat{x}$ as $m \to \infty$.

It is also possible to prove convergence rates under appropriate assumptions on the data and in fact the optimal asymptotic order of accuracy $O(\max(\delta^{2/3},\tau))$ is attainable. For details and numerical illustrations see Callon [2].

REFERENCES

1. K. E. Bullen, "The Earth's Density", Chapman and Hall, London, 1975.

2. G. D. Callon, Numerical regularization for restricted pseudosolutions, in "Approximation Theory V", (C. K. Chui, L. L. Schumaker, and J. D. Ward, Eds.), Academic Press, New York, to appear.

3. G. D. Callon and C. W. Groetsch, The method of weighting and approximation of restricted pseudosolutions, J. Approximation Th., to appear.

4. L. Eldén, A weighted pseudoinverse, generalized singular values, and constrained least squares problems, BIT 22(1982), 487-502.

5. C. W. Groetsch, "Generalized Inverses of Linear Operators: Representation and Approximation", Dekker, New York, 1977.

6. C. W. Groetsch, "The Theory of Tikhonov Regularization for Fredholm Equations of the First Kind", Pitman, London, 1984.

7. J. N. Holt and A. J. Bracken, First-kind Fredholm integral equation of liver kinetics: numerical solutions by constrained least squares, Math. Biosciences 51(1980), 11-24.

8. J. Locker and P. M. Prenter, Regularization with differential operators I: general theory, J. Math. Anal. Appl. 19(1980), 504-529.

9. N. Minamide and K. Nakamura, A restricted pseudoinverse and its application to constrained minima, SIAM J. Appl. Math. 19(1970), 167-177.

10. V. A. Morozov, "Methods for Solving Incorrectly Posed Problems", Springer-Verlag, New York, 1984.

Department of Mathematical Sciences
University of Cincinnati
Cincinnati, Ohio 45221-0025
U.S.A.

ON THE INVERSION OF SOME FIRST-KIND FREDHOLM EQUATIONS OCCURRING IN OPTICAL APPLICATIONS

E R Pike

Dept of Physics, King's College, London, UK
and
Centre for Theoretical Studies, RSRE, Malvern, Worcestershire, UK

1. INTRODUCTION

My interest in the inversion of first-kind Fredholm equations was first aroused seriously when I was required to do something about the so-called polydispersity problem. This comes about when using laser light scattering to measure the Brownian motion of macromolecules and hence, essentially, to measure their size. The spectrum of intensity fluctuations of light scattered by a suspension of identical particles can be shown to be of Lorentzian form with half-width proportional to the diffusion constant of the molecules in the suspension. A practical technique, now widely used, measures, by special high-speed digital hardware, the autocorrelation function of the digital stream of detected single-photon events (photon correlation spectroscopy [1], or PCS), which is related by the Wiener-Khintchine theorem to the spectrum and thus gives rise to an exponential function of decay constant proportional to the inverse of the diffusion constant. In a typical experiment, 10^7 or more scattered single photons will be analysed using digital circuitry and a digital read-out of the exponential correlation function will be available with accuracy of the order of 0.1% on each point. These points are chosen to cover a few decay times in either a linear or more preferably an integrated logarithmic sampling scheme. The data reduction problem in this "monodisperse" case, namely, fitting a single exponential curve, poses no serious problem. When particles of a number of different molecular sizes, however, are present - the "polydisperse" case - each size fraction gives rise to its own exponential curve and hence the total scattering produces an autocorrelation function which is a sum of exponentials and the data reduction problem becomes one of Laplace inversion.

To one tender in the art of inverse problems the Laplace inversion problem comes as something of a surprise. A simple first approach [2] is to attempt to find the moments or cumulants of the required size distribution. Thus, we start with the equation for the sampled data,

$$g(y_i) = \int_0^\infty e^{-xy_i} f(x)\, dx \qquad i = 1, \ldots N \tag{1}$$

where x is some measure of the particle size and f(x) is the required polydispersity density function. Taking the logarithm and expanding in $x - \bar{x}$, we obtain

$$\log g(y_i) = \sum_{n=0}^{\infty} \frac{(-1)^n}{n!} c_n (\bar{x} y_i)^n \tag{2}$$

where

$$c_m = \int_0^\infty \frac{(x-\bar{x})^m}{\bar{x}^m} f(x)\, dx \tag{3}$$

are, approximately, the cumulants of the distribution.

A logarithmic plot of the data thus has a Taylor expansion about the origin on the positive real line which has the cumulants as successive coefficients. Polynomial fitting to nth order will thus give the first n cumulants.

What one finds in practice is that in spite of feeding, say, 100 numbers, each of accuracy 0.1%, into the calculation, the only moment which can be determined with any real accuracy is the first! The errors on the second moment are such that it gives only a very qualitative measure of the polydispersity and the third moment is not used. The idea of reconstructing a distribution curve, f(x), for the particle sizes was, for a long time, not seriously pursued.

To try to understand this problem more fully we went back to the work of Slepian, Landau and Pollak on another ill-conditioned Fredholm equation of the first kind, namely, that describing diffraction limited imaging or band-limited communication, namely,

$$g(y) = \frac{1}{2\pi}\int_{-\Omega}^{\Omega}\Big[e^{-iky}\int_{-X/2}^{X/2} e^{ikx} f(x)\,dx\Big]\,dk = \int_{-X/2}^{X/2} \frac{\sin\Omega(y-x)}{\pi(y-x)} f(x)\,dx \qquad (4)$$

This equation was solved by these authors by an eigenfunction expansion (by use of the first "miracle" described by Grunbaum elsewhere in this volume)

$$f(x) = \sum_{n=0}^{\infty} \frac{1}{\lambda_n}\Big[\int_{-X/2}^{X/2} g(y)\,\Phi_n(y)\,dy\Big]\,\Phi_n(x) = \sum_{n=0}^{\infty} \frac{(g.\Phi_n)}{\lambda_n}\,\Phi_n \qquad (5)$$

where

$$\lambda_n \Phi_n(y) = \int_{-X/2}^{X/2} \frac{\sin\Omega(y-x)}{\pi(y-x)}\,\Phi_n(x)\,dx \qquad (6)$$

and thus Φ_n and λ_n are the eigenfunctions and eigenvalues, respectively, of the integral operator K from $L^2(-X/2,X/2)$ to $L^2(-X/2,X/2)$ described by

$$(Kf)(y) = \int_{-X/2}^{X/2} \frac{\sin\Omega(y-x)}{\pi(y-x)}\,f(x)\,dx \qquad -\frac{X}{2} \le y < \frac{X}{2} \qquad (7)$$

The scalar product is denoted (*,*).

These Φ_n are called prolate spheroidal functions and the remarkable form of their eigenvalue spectrum is well known, being essentially unity up to a value of $n = X\Omega/\pi$, the "Shannon number", S , and then dropping suddenly to very low values for higher n. They are complete and orthonomal in $L^2(-X/2,X/2)$. The experimental consequences of this method of solution are that if the function f(x) is decomposed into its eigenfunction components then those beyond the Shannon number will, to all intents and purposes, have to be lost in any inversion scheme since they are "transmitted" by the instrument so weakly that they will be dominated by the inevitable noise present in real experiments and will be unable to be recovered. The summation (5) thus needs to be terminated at n=S and the reconstrunction considered necessarily only as a projection of the function f on the subspace spanned by the first S eigenfunctions. If higher order components were present in f then the only possibility for recovering any information about them is that some "model" might exist for f which connects the coefficients of the "visible" and "invisible" components in some way. No explicit examples of this possibility, however, come to mind. The result of the truncation of the summation can be visualised by considering the total effect of both the transmission by the instrument and the subsequent recovery procedure in terms of an "impulse response function", $S(x, x_0)$, or averaging operator, which is the reconstruction achieved when f is a delta function at x_0. We have in this case

$$(g, \varphi_n) = \alpha (f, \varphi_n) = \alpha \varphi_n(x_0) \tag{8}$$

and thus

$$S(x, x_0) = \sum_{n=0}^{S} \varphi_n(x_0)\, \varphi_n(x) \tag{9}$$

This function may be computed for given values of S determined by the space-bandwidth product, $X\Omega/\pi$, of the problem at hand. It has the form of a central lobe flanked by decreasing side lobes, the width of the central lobe decreasing to zero as the value of S tends to infinity. In this and the more general cases to be discussed later this impulse response function may be used to define the "resolution" or number of degrees of freedom of the instrument as a function of x_0 and S or other analogous parameters. In this particular case, due to the sharp cut off of the spectrum at S, the number of "visible" components is essentially independent of the noise level and the impulse response function computed in the manner indicated above has the basic form of the kernel of the equation, $[\sin\Omega(x-x_0)]/\pi(x-x_0)$. In equation (5) a hard truncation has been applied, but various "softer" roll-off conditions ("windows") may be used to regularize the problem as desired.

This method of eigenfunction expansions will be applied to the Laplace transform inversion, to go back to our original problem, and also to other first-kind Fredholm equations occuring in optics. In these, the spectrum will, of course, not have the above special form and the generalised Shannon number or the index of the eigenvalue at which the "image component" is judged to have been reduced to the noise level, will vary with the noise level, as will, therefore, the resolution or width of the averaging operator in each individual case.

2. THE LAPLACE TRANSFORM

In 1978 McWhirter and Pike [3] applied the above method of eigenfunction expansions to the Laplace inversion and found explicit eigenfunctions and eigenvalues not only for this problem, but for all first-kind Fredholm equations of the form

$$g(y) = \int_0^\infty K(xy)\, f(x)\, dx \qquad 0 \le y \le \infty \tag{10}$$

These are

$$\varphi_\omega^{\pm}(x) = \begin{matrix} Re \\ Im \end{matrix} \left\{ \frac{[\sqrt{\tilde{K}(\frac{1}{2}+i\omega)}\; x^{-\frac{1}{2}-i\omega}]}{\sqrt{(\pi\, |\tilde{K}(\frac{1}{2}+i\omega)|)}} \right\} \tag{11}$$

$$\lambda_\omega^{\pm} = \pm\, |\tilde{K}(\tfrac{1}{2}+i\omega)| \tag{12}$$

Where $\tilde{K}$ is the Mellin transform of K. The exponential fall off of the eigenvalues gives rise to an averaging operator composed of the sum of very few terms which is thus very broad. This is the reason for the difficulties which one finds in practice with Laplace transform inversion.

A number of other optical problems can be approached in this way using the above results of McWhirter & Pike. These solutions will be outlined only briefly in the following sections, since we shall proceed later in the paper to more precise, but related methods.

3. FRAUNHOFER SCATTERING

The scattered intensity versus angle near the forward direction from a suspension of opaque spheres has the form of the first-kind Fredholm equation (10) with

$$K(z) = \frac{J_1^2(z)}{z^2} \qquad (13)$$

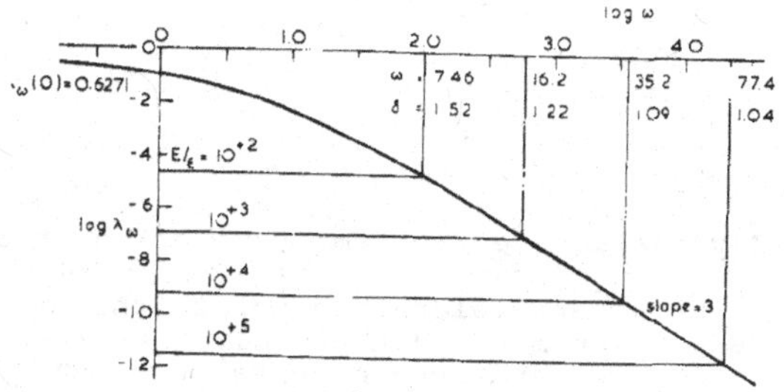

Fig 1 Eigenvalue spectrum for Fraunhofer scattering.

and where y is the scattering vector and x the particle radius. The measured intensity distribution can thus be used to determine particle size distributions f(x). This problem may be solved by eigenfunction expansions, as above. In Fig 1 we show the eigenvalue spectrum for this problem. The fall off in this case is as x^3, so the problem is not as badly conditioned as the Laplace inversion and, in general, a sufficient number of parameters may be found from an inversion to reconstruct reasonable distributions. If the spheres are not completely opaque, the Fraunhofer kernel of equation (13) is only an approximation and then either a full Mie scattering kernel or another approximation like the anomalous scattering approximation must be used instead.

4. EXTINCTION

This is a similar problem to the Fraunhofer scattering problem of the previous section. A spherical particle suspension is illuminated by a plane wave, usually a collimated light beam, and the reduction in intensity at the zero degree forward angle is measured as a ratio to the intensity when the particles are not present. This is repeated for a number of different wavelengths. In the polydisperse case this experiment gives rise to the first-kind Fredholm equation (10) with

$$K(z) = 3\left(2 + \frac{4}{z} - \frac{4\sin z}{z} - \frac{4\cos z}{z^2}\right)/4z \tag{14}$$

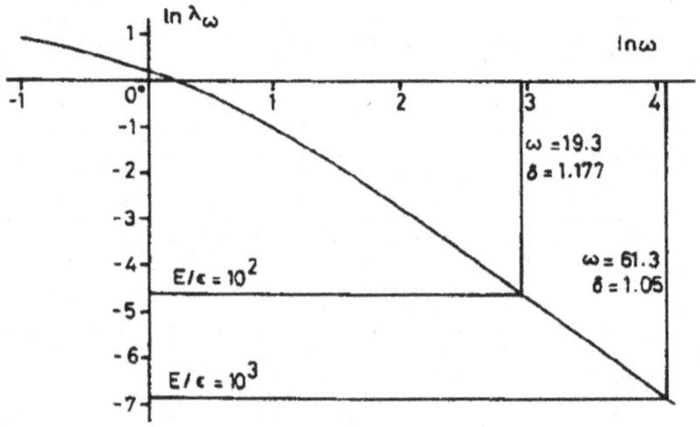

Fig 2 Eigenvalue spectrum for Extinction problem.

and where now y is the wavelength and x the particle radius. Theoretically this problem may again be solved by the eigenfunction expansion method and we show the eigenvalue spectrum in Fig 2. However, the limited range of wavelengths normally available is a difficulty and the more powerful methods of singular system expansions to be introduced below are more suitable.

5. LASER DOPPLER VELOCIMETRY

The method of spectral analysis of scattered light in which the digital correlation function of a photon stream is formed, described in the Introduction (PCS), can be used with advantage to measure the velocity of gas and liquid flows and also the motion of solid bodies. In the first two cases dust paricles naturally occuring in the flow normally provide sufficient scattering centres although sometimes extra particles may be added. The passage of a scattering centre at a uniform velocity through a laser beam gives rise to a Doppler-shifted scattering, which may be mixed on a nonlinear detector with unshifted light or light shifted by a fixed known amount, or, most often, with light scattered from a second laser beam of the same frequency but incident at a different angle. In all cases a low frequency beat is apparent in the rate of photon detections and this appears in the photon correlation function. This function also decays on the time scale of passage of the particles through the laser beam which normally can be taken to have a Gaussian intensity profile. The velocity may be recovered by fitting the data to a function of the form

$$K(z) = e^{-z^2/2z_0^2}(1 + m\cos sz) \tag{15}$$

Here z_0 represents the laser beam diameter, m is a visibility factor and s gives the frequency scale of the Doppler shift.

When the data is accumulated over a period in which the velocity varies or in which turbulence occurs, the distribition of velocities or turbulence intensity may be derived by inversion of an equation of the form (10) with the kernel given above. In this case, y is the time delay variable of the correlation function and x the velocity of the medium.

The ill-conditioned nature of this inversion depends on the ratio of the frequency scale parameter s to the beam width z_0. This gives the number of oscillations of the cosine term within the damping imposed by the Gaussian profile. If many such oscillations are available the problem approaches the Fourier transform, the inversion is not then badly conditioned and spline models have been used successfully [4]. For high-speed flows, however, electronic limitations force us into the situation in which only a few cycles of oscillation are available and work is in hand to apply the eigenfunction expansion method in this case. McWhirter and Pike [3] solved a similar problem where the damping profile was a pure exponential rather than Gaussian, viz

$$K(z) = e^{-\alpha z}\cos\beta z \tag{16}$$

and found the eigenvalue spectrum

$$|\lambda_\omega^\pm|^2 = \frac{\Gamma(\frac{1}{2} + i\omega)\cos[(\frac{1}{2} + i\omega)\tan^{-1}(\beta/\alpha)]}{(\alpha^2+\beta^2)^{\frac{1}{4} + \frac{i\omega}{8}}} \tag{17}$$

where

$$\tau = \frac{\alpha}{\beta} \tag{18}$$

is the number of effective oscillations discussed above. The dependence on this parameter of the rate of fall off of the spectrum and hence of the number of terms which contribute to the impulse response function can be clearly seen.

6. EXPONENTIAL SAMPLING

The work of McWhirter and Pike on the Laplace transform led Ostrowsky et al [5] to consider a sampling scheme for a model inversion in which the reconstruction was attempted with a resolution decreasing exponentially with increase of radius parameter. This is the equivalent of the Nyquist sampling of Fourier theory when the problem is dilationally rather than translationally invariant.

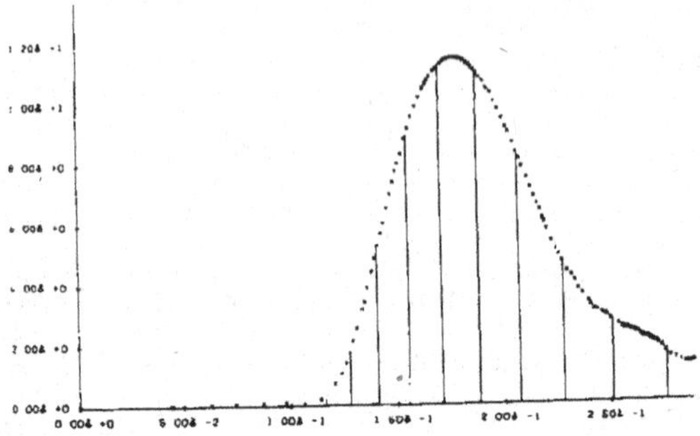

Fig 3 "Exponential sampling" inversion of a polydisperse distribution of silver halide particles using 11 sample points.

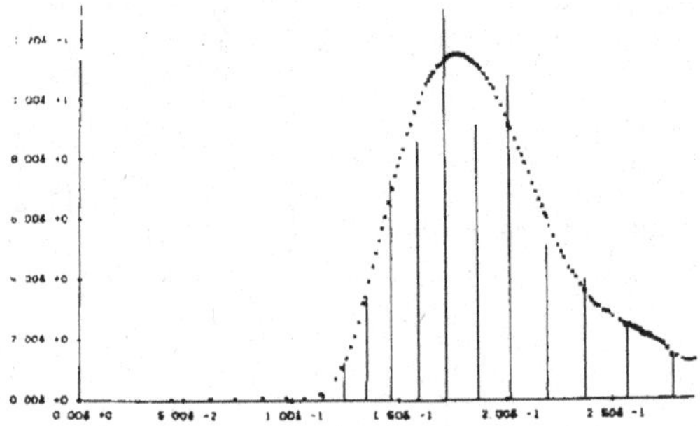

Fig 4 As Fig 3 but with 12 sampling points.

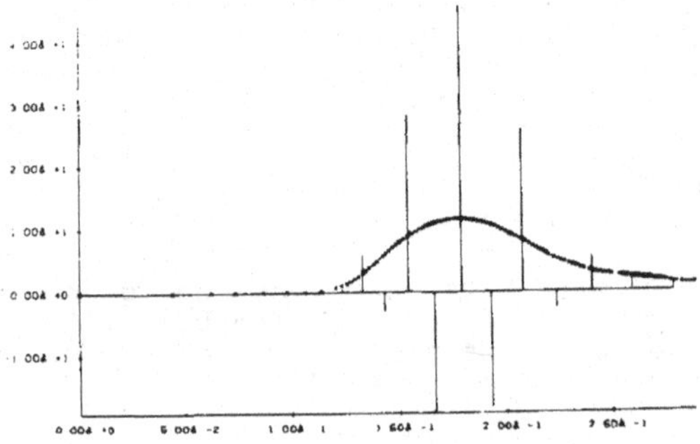

Fig 5 As Fig 3 but with 13 sampling points.

Matrix inversions based on this so called "exponential sampling" method using data provided by Kodak research laboratories representing a realistic problem of particle sizing in the photographic industry are show in Figs 3-5. The ill-conditioned nature of the inversion is shown up by the onset of instability when the sampling rate becomes too high compared with the level of round-off error of the computer.

As experience with this exponential sampling method built up it became apparent that the resolution achievable, although not high, was significantly better than that predicted by the eigenvalue spectrum. Interpolation between the sampled points of the reconstruction can be made using a sinc function on a logarithmic scale in a manner quite analogous to that of sampling theory in the Fourier case. Versions of this method are still in use in some commercial instruments although, as we shall see below, further developments have elucidated the resolution problem and have led to more accurate and efficient inversion techniques.

7. SINGULAR FUNCTION ANALYSIS, FINITE LAPLACE TRANSFORM

The origin of the discrepancies in resolution described in the previous section where shown by Bertero et al [6] to be due to an implicit restriction of the support of the reconstruction of the solution in the matrix inversion. Whereas the eigenvalue spectrum is calculated for a possible infinite support, in practice finite limits must be set for numerical calculations and these, of necessity, add some a priori knowledge of the position and extent of the solution. This knowledge can be used explicity to calculate the extra resolution possible by an extension of the method of eigenfunction expansions known as singular function expansions, well known in pure mathematics since the last century [7] but little used in physics until quite recently.

If the support of the unknown solution f(x) is known to be restricted to the interval [a, b] then we have the finite Laplace inversion problem

$$g(y) = \int_a^b e^{-xy} f(x)\, dx \qquad 0 \leq y \leq \infty \tag{19}$$

which is beyond the scope of the normal eigenvalue expansion method since g and f are functions which lie in different functional spaces, say, Y and X, respectively. We may suppose that these may be taken to be Hilbert spaces for most cases of application and, in particular, for the finite Laplace inversion we shall consider for the moment Y to be $L^2(0,\infty)$ and X, $L^{(2)}(a,b)$. The mapping

$$X \ni f \rightarrow g \in Y \tag{20}$$

described by equation (19) can be denoted by a linear operator $K:X \rightarrow Y$ which is onto in this case. The singular function analysis now requires a set of solutions $u \in X$ and $v \in Y$ s.t.

$$K u_n = \alpha_n v_n \tag{21}$$

$$n = 0, 1 \ldots$$

$$K^* v_n = \alpha_n u_n \tag{22}$$

where the α_n are real singular values and the star denotes the adjoint operator defined by

$$(K^*g)(x) = \int_0^\infty e^{-xy} g(y)\,dy \qquad a \le x \le b \tag{23}$$

It is readily seen that the u_n's are eigenfunctions of the integral operator $K^*K: X \to X$, with eigenvalues α_n^2, while the v_n's are eigenfunctions of the integral operation $KK^*: Y \to Y$, with the same eigenfunctions. We shall call the triple $(\alpha_n; u_n, v_n)_{n=0}^{\infty}$, the <u>singular system</u> of the operator K.

The formal solution of the inversion problem in terms of the singular system is

$$f(x) = \sum_{n=0}^{\infty} \frac{(g, v_n)_Y}{\alpha_n} u_n(x) \tag{24}$$

which is a simple extension of the inversion formula (5). $(*,*)_Y$ denotes the scalar product in the Hilbert space Y.

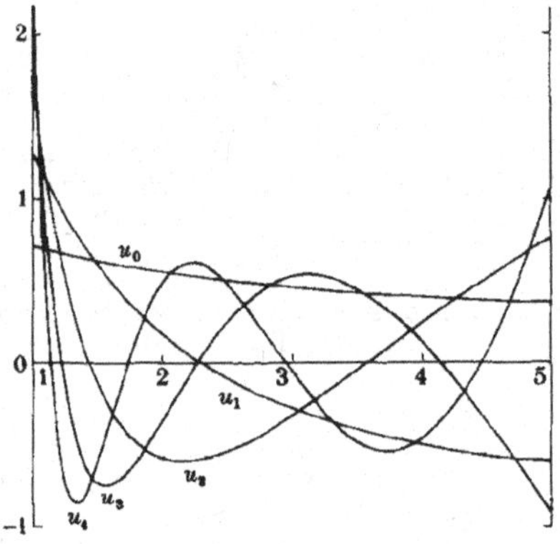

Fig 6 Low-order singular functions of the finite Laplace inversion problem.

In the work of Bertero et al on the finite Laplace transform [6] it was shown that the support of the solution can be taken to be $[1,\gamma]$ in suitable normalised coordinates without loss of generality and the first few singular functions u_n are shown in Fig 6 for a value of γ of 5. Numerical calculations of the eigenfunctions of K*K were made using the method of tensor products of splines due to Hammerlin and Schumaker [8] followed by the power method and deflation. Assuming a "white noise" spectrum for both signal and noise components Bertero and Pike computed the resolution which can be achieved in inversions with various levels of "signal-to-noise" ratio as a function of γ. These are shown in Fig 7. As a rule of thumb for the Brownian motion experiments with $2 < \gamma < 5$ and signal to noise ratio ~ 10^3 it may be seen that whatever <u>a priori</u> suport is used within the above range there will be of the order of three significant terms in the reconstruction expansion. Since in this case the order of each term gives the number of nodes of the corresponding singular function it is seen that the resolution which may be achieved improves roughly linearly with finer a priori specification of the support. The impulse response function is shown in Fig 8 for various values of x_o and shows the dilational loss of resolution as x_o increased. The original paper should be consulted for more details.

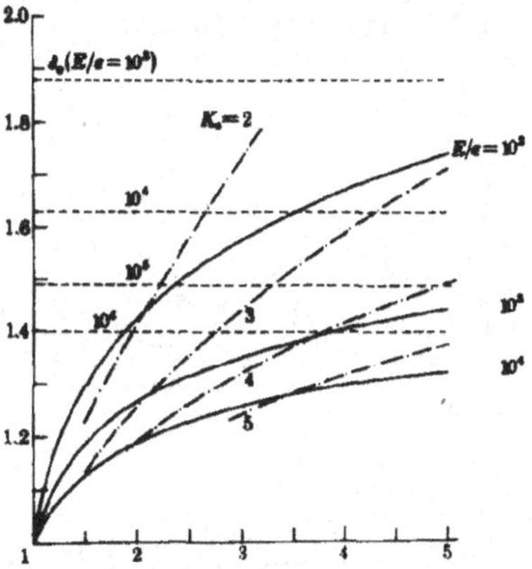

Fig 7 Resolution ratio versus support for the finite Laplace transform for various values of signal-to-noise ratio.

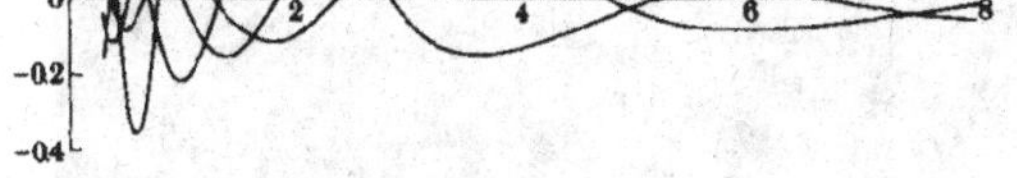

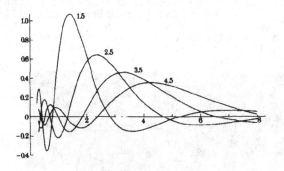

Fig 8 Impulse response function for the finite Laplace transform for $S/N \sim 10^3$ for various values of x.

8. SINGULAR FUNCTION ANALYSIS, DIFFRACTION LIMITED IMAGING.

We were originally led to consider the eigenfunction expansion method by reference to the diffraction limited imaging problem or the band-limited Fourier transform. The analogous resolution limits were, however, improved by introducing the more powerful method of singular value analysis which allowed a priori support constraints to be used. It is natural to ask if the original problem would benefit in the same way by the application of this more general theory and this question was discussed by Bertero & Pike in 1982 [9]. Improvement in this case over the usual eigenfunction analysis limit, the classical Rayleigh resolution, tends to be called "superresolution". In the original terms of the problem this, of course, is not possible but, regarded as a singular value problem with different a priori supports for object and image the classical analysis of resolution does not apply and by measuring the complete image we may superresolve in practice to any degree desired depending on the space-bandwidth product. In Fig 9 we give the superresolution gain as a function of this parameter for various signal to noise ratios for square and circular pupils taken from the work of these authors.

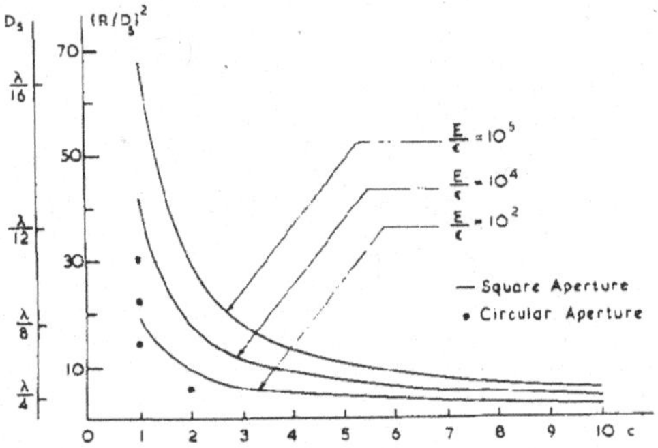

Fig 9 Superresolution gain as a function of space-bandwidth product for various values of S/N ratio. The solid lines are for a square pupil and the dots are for a circular pupil.

In Figs 10 and 11 we show a practical demonstration where a singular function expansion has been used to resolve detail well beyond the Rayleigh limit taken from the work of Walker [10] and, with permission, from unpublished calculations of Greenaway. A similar superresolution gain was achieved with the same data, in the paper of Walker cited, by an iterative technique.

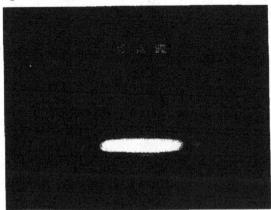

Fig 10 Object for superresolution experiment and corresponding unresolved image.

We have come to call the above type of singular system analysis the continuous-continuous case with a priori top-hat support. The method assumes not only that we search for a finitely supported L^2 function for the solution, but that the data is provided also, as such a continuous function at least in the Laplace inversion over an essentially infinite support. This latter assumption is clearly far from the truth in many cases and we have taken advantage in further developments of the fact that, since we are mapping between different spaces in the singular system method, we may take the "data space", Y, to be a Euclidean vector space rather than $L^2(0,\infty)$.

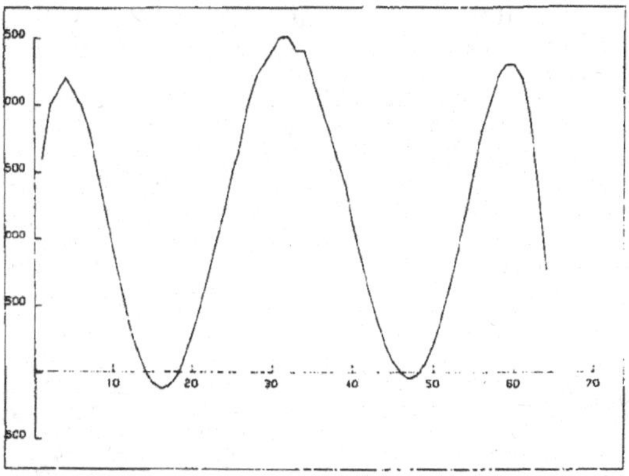

Fig 11 Reconstruction of object of Fig 10 by singular function expansion to 9 terms.

9. SINGULAR SYSTEM ANALYSIS, CONTINUOUS-DISCRETE CASE

Data in most experiments is, or can easily be, presented in the form of truncated and sampled values of the instrumental output. The sampling may be linear, logarithmic or nonlinear in some other way and the truncation may be severe, as in the extinction problem, or of less importance as we have seen in the diffraction limited imaging problem. In any case the data can be well represented by a Euclidean vector in an M-dimensional vector space where M is the number of data points registered. The direct or forward problem of calculating the data vector given the object function, still assumed to be an L^2 function in the Hilbert space X, is a mapping

$$K:\ X \ni f \rightarrow g \in Y \tag{25}$$

and the inversion can be considered as before by solving the system

$$K^{*}K\,u_n = \alpha^2 u_n \tag{26}$$

$$K K^{*} v_n = \alpha^2 v_n \tag{27}$$

where K*K is a finite rank integral operator and KK* an MxM matrix.

It is normally easier to find the N, say, (N < M) singular system components required to form the impulse response function by applying the power method and deflation to KK* to give the first N eigenvectors and eigenvalues. The corresponding u_n are then found from

$$\alpha_n u_n = K^* v_n \tag{28}$$

More sophisticated SVD methods such as the Gollub-Reinsch algorithm (see F Natterer, this volume) are rarely appropriate for experimental (as distinct from numerical) work since they are more space and time consuming and are designed to obviate ill conditioning problems which have already been solved by the series expansion truncation dictated by experimental noise levels. The only real exception that comes to mind to this observation is when dedicated data reduction hardware needs to be used so that numerical round-off errors increase to the order of experimental noise levels (F Natterer, private communication).

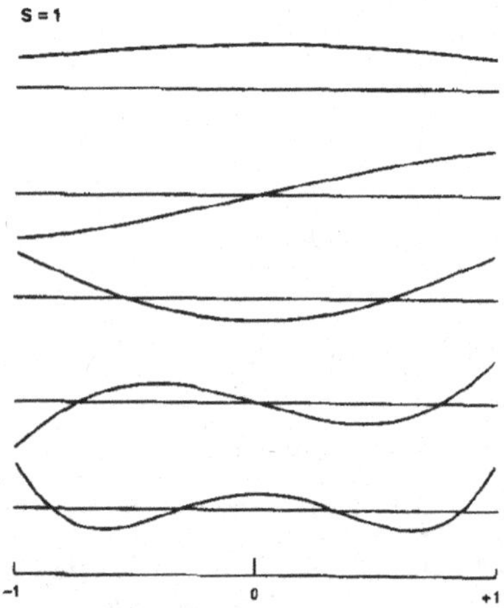

Fig 12 The first five singular functions for the diffraction-limited imaging problem with 21 discrete data points. S=1.

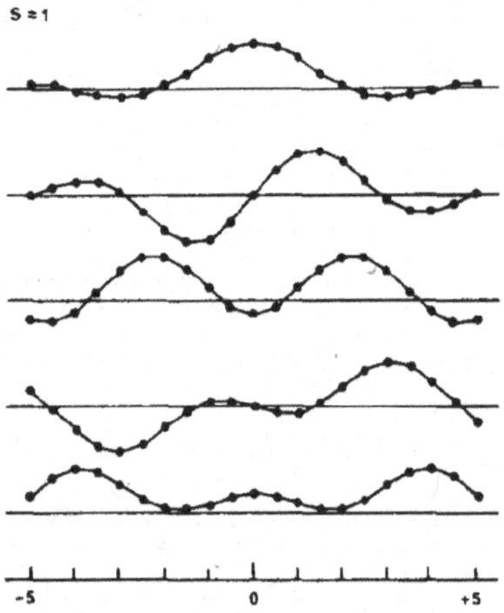

Fig 13 The first five singular vectors corresponding to the singular functions of Fig 12.

In Figs 12 and 13 we show the first five singular functions and the first five singular vectors for the diffraction limied imaging problem with $X\Omega/\pi = 1$. The data has been truncated at $\pm$ 5 times the geometrical image and 21 equally spaced data points have been used.

10. WEIGHTED SPACE RECONSTRUCTIONS

One further refinement takes us to the methods which we use today for these experimental data inversion problems. This arose from the difficulty that the singular functions are usually not small at the boundaries of their domains of support and thus truncated or regularised expansions show a type of "Gibbs phenomenon" in that undesirable ringing phenomena appear at these boundaries in the reconstructions. The remedy is to modify the "top-hat" nature of the a priori assumed support by using weighted L^2 space reconstructions so that the singular functions are forced to be small in these exterior regions. We can look, therefore, for functions in the class

$$\int_0^\infty \frac{|f(y)|^2}{|P(y)|^2}\,dy < +\infty \qquad (29)$$

where P(x) is a "profile function" expressing either a priori statistical knowledge of the approximate position and location of the object or, alternatively, P(x) may express other localisation conditions, for example, it can be used to describe illumination conditions in microscopy.

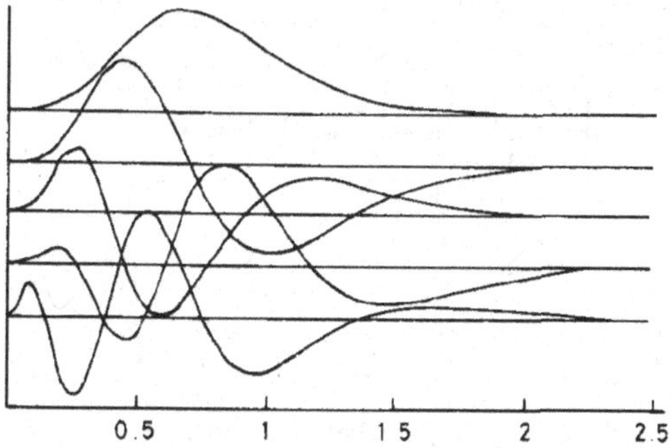

Fig 14 The low order singular functions for the finite Laplace transform with a Γ-function as weighting profile function.

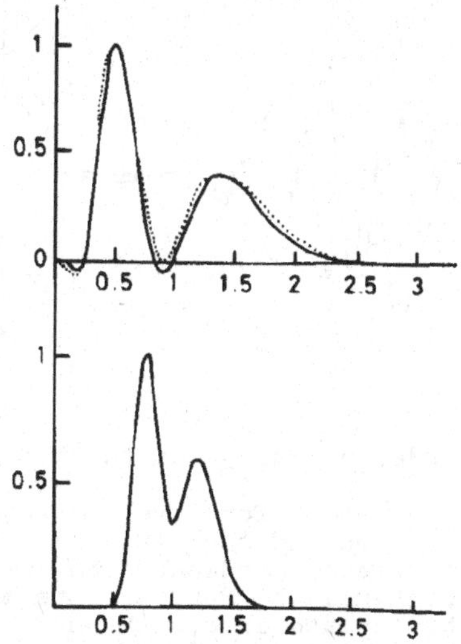

Fig 15 Reconstruction of a bimodal distribution from its finite and sampled Laplace transform using five terms of its singular function expansion.

The theory of singular systems in such weighted spaces is given in detail in the original papers [11,12] and in the lectures of Bertero in this volume and we shall be content with giving examples of applications to some of the optical problems described above. In Fig 14 we show the low order singular functions for the finite Laplace inversion where a Γ-distribution of mean and variance matching the experimental data (determined independently by the cumulants method) is used as a profile function. These are found to be very similar and have similar singular values for the two cases of 64 linearly spaced data points and for 5 logarithmically spaced data points over the same span of delay values. Fig 15 shows a reconstruction of a bimodal distribution using 5 terms.

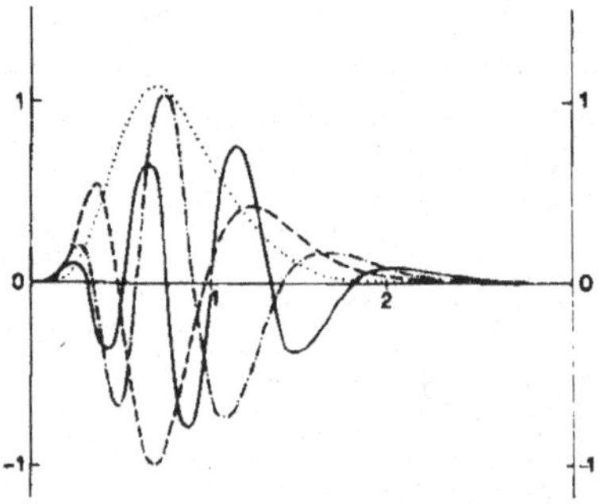

Fig 16 Low order even singular functions for the Fraunhofer diffraction problem with γ = 5 and Γ-function weighting profile.

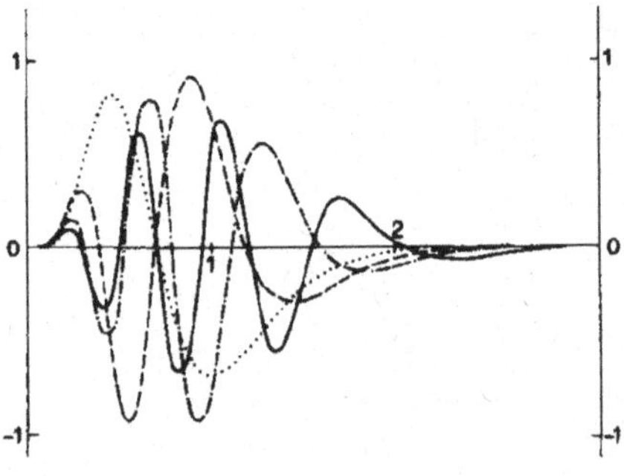

Fig 17 Low order odd singular functions as Fig 16.

Calculations have been made for the Fraunhofer diffraction and extinction problems and even and odd singular functions are shown for these problems with parameters indicated in the captions in Figs 16, 17, 18, and 19 respectively. Many more terms may be used in these problems than in the Laplace inversion and in Fig 20 is shown a reconstruction of a six-peaked particle size distribution using 23 terms of the appropriate singular function expansion. It should be said that the singular system is usually precomputed and held in store, so that the process of computation

$$f(x_i) = \sum_{m=0}^{M-1} \sum_{n=0}^{N-1} \left(g_m \cdot \frac{v_{nm}}{\alpha_n} \right) u_n(x_i) \qquad i = 1, \ldots L \tag{30}$$

would seem to involve the minimum number of operations taking M data points into L object points with N degrees of freedom. This computational load for a general inversion problem can only be reduced by utilising any symmetry properties of the matrix elements.

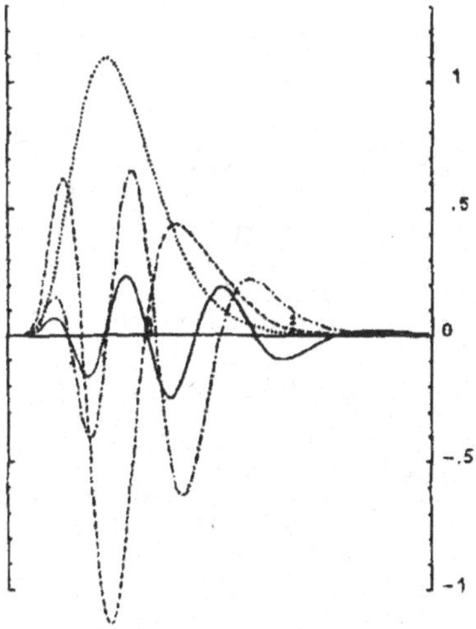

Fig 18 Low order even singular functions for the extinction problem with $\gamma = 5$ and Γ-function weighting profile.

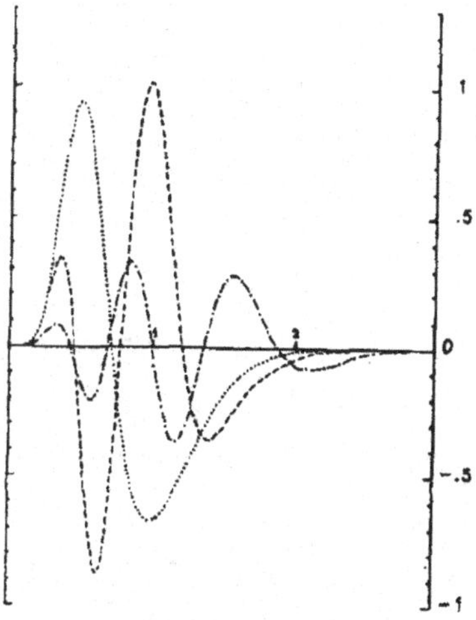

Fig 19 Low order odd singular functions as Fig 18.

11. CONFOCAL SCANNING MICROSCOPY

The weighted space approach can be used very effectively in confocal scanning microscopy. In a confocal scanning laser microscope an illuminating parallel laser beam is focussed on the specimen by a high aperture "illumination" lens and the transmitted light is collected by a similar lens to form an image. This is the situation described above in Section 8 save that the a priori "top-hat" support of that problem is replaced by a profile function representing the incident illumination. This problem has been considered by Bertero et al [13] and singular systems calculated for various numbers of detectors in one and two dimensions. In Fig (21) we show a simulated amplitude transmission object, together with images formed by an ordinary microscope, a normal confocal microscope in which a single detecting element is used at the centre of the image, and a reconstructed object from a singular function summation of 5 terms. An analytical treatment of this type of problem has been given by Gori and Guattari [14] for the case of a sampled but non-truncated image.

(a)

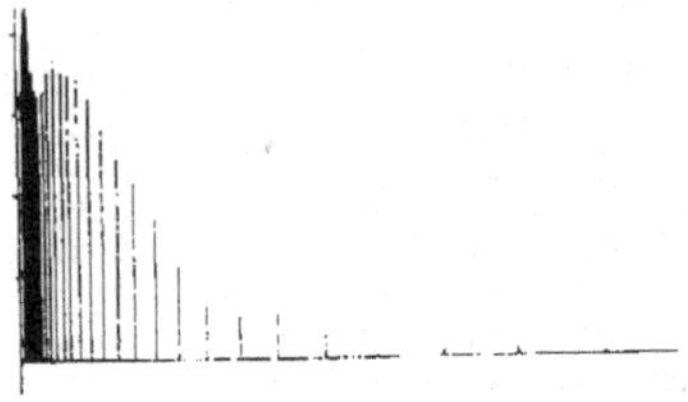

(b)

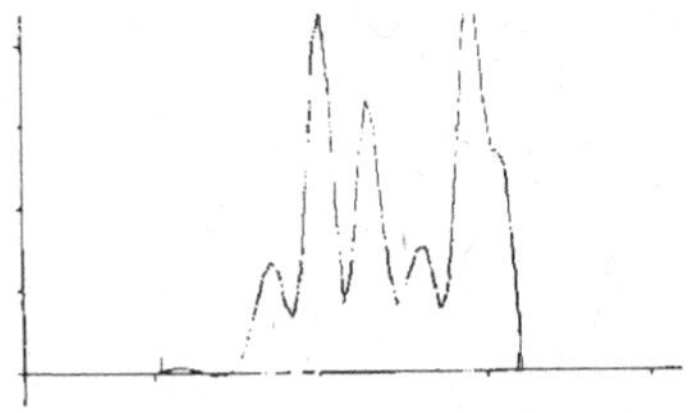

Fig 20 Reconstruction of a six-peaked particle size distribution from its Fraunhofer diffraction pattern, using 23 terms of its singular function expansion.

(a) Intensity v scattering wave vector K.

(b) Recovered particle size distribution versus radius.

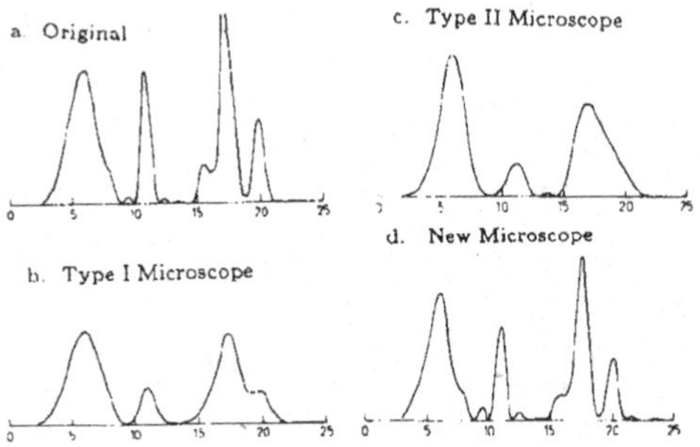

Fig 21 Simulated amplitude object (original) (a), with images from Type I microscope (normal microscopy) (b), Type II microscope (normal confocal scanning microscopy) (c), and reconstructed object (d) by singular function expansion.

12. CONCLUSIONS

Recent work on the inversion of some first-kind Fredholm equations occuring in optical applications have been surveyed and a rather universal approach has been evolved in which the experiment is regarded as a continuous to discrete mapping from the object to the measured data. The mapping is inverted by decomposition into orthnormal bases in the continuous object (Hilbert) space and the discrete data (vector) space. Suitable weights are applied in both object and data spaces to account for *a priori* or experimental constraints and the object reconstruction is made by a suitable truncated singular system expansion or by replacing the truncation by suitable window function applied to the singular spectrum. The method is linear and computationally efficient. Positivity constraints can also be incorporated if desired (see Bertero, this volume).

13. ACKNOWLEDGEMENTS

This work has been performed over recent years in collaboration with Dr De Mol and colleagues at the University of Brussels, Dr Brakenhoff and colleagues at the University of Amsterdam and Professor Bertero and colleagues at the University of Genoa, as well as with colleagues at Malvern, in particular Dr J McWhirter and Dr G de Villiers. We acknowledge gratefully assistance from NATO Grant No 463/84 and EEC Grant No ST2J-0089-1-UK (CD).

REFERENCES

[1] Cummins H Z and Pike E R (eds), 1977, Photon Correlation Spectroscopy and Velocimetry (Plenum, New York)

[2] Pusey P, 1974, In Photon Correlation & Light Beating Spectroscopy edited by H Z Cummins and E R Pike (Plenum, New York)

[3] McWhirter J G and Pike E R, 1978, J Phys A 11 1729.

[4] McWhirter J G, 1980, Optical Acta, 27, 83.

[5] Ostrowsky N, Sornette D, Parker P and Pike E R, 1981, Optica Acta 28, 1059.

[6] Bertero M, Boccacci P and Pike E R, 1982, Proc Roy Soc Lond, A383, 15.

[7] Picard E, 1910, RC Mat Palermo, 29, 615.

[8] Hämmerlin G and Schumaker L L, 1980, Numer Math 34, 125.

[9] Bertero M and Pike E R, 1982, Optica Acta, 29, 727.

[10] Walker J G, 1983, Optica Acta, 30, 1197.

[11] Bertero M, Brianzi, P and Pike E R, 1985, Inverse Problems, 1, 1.

[12] Bertero M, De Mol C and Pike E R, 1985, Inverse Problems, 1, 301.

[13] Bertero M, De Mol C, Pike E R and Walker J G, 1984, Optica Acta 31, 923.

[14] Gori F and Guattari G, 1985, Inverse Problems, 1, 67.

LIST OF C.I.M.E. SEMINARS

		Publisher
1954 -	1. Analisi funzionale	C.I.M.E.
	2. Quadrature delle superficie e questioni connesse	"
	3. Equazioni differenziali non lineari	"
1955 -	4. Teorema di Riemann-Roch e questioni connesse	"
	5. Teoria dei numeri	"
	6. Topologia	"
	7. Teorie non linearizzate in elasticità, idrodinamica, aerodinamica	"
	8. Geometria proiettivo-differenziale	"
1956 -	9. Equazioni alle derivate parziali a caratteristiche reali	"
	10. Propagazione delle onde elettromagnetiche	"
	11. Teoria della funzioni di più variabili complesse e delle funzioni automorfe	"
1957 -	12. Geometria aritmetica e algebrica (2 vol.)	"
	13. Integrali singolari e questioni connesse	"
	14. Teoria della turbolenza (2 vol.)	"
1958 -	15. Vedute e problemi attuali in relatività generale	"
	16. Problemi di geometria differenziale in grande	"
	17. Il principio di minimo e le sue applicazioni alle equazioni funzionali	"
1959 -	18. Induzione e statistica	"
	19. Teoria algebrica dei meccanismi automatici (2 vol.)	"
	20. Gruppi, anelli di Lie e teoria della coomologia	"
1960 -	21. Sistemi dinamici e teoremi ergodici	"
	22. Forme differenziali e loro integrali	"
1961 -	23. Geometria del calcolo delle variazioni (2 vol.)	"
	24. Teoria delle distribuzioni	"
	25. Onde superficiali	"
1962 -	26. Topologia differenziale	"
	27. Autovalori e autosoluzioni	"
	28. Magnetofluidodinamica	"
1963 -	29. Equazioni differenziali astratte	"
	30. Funzioni e varietà complesse	"
	31. Proprietà di media e teoremi di confronto in Fisica Matematica	"

1964 -	32. Relatività generale	C.I.M.E.
	33. Dinamica dei gas rarefatti	"
	34. Alcune questioni di analisi numerica	"
	35. Equazioni differenziali non lineari	"
1965 -	36. Non-linear continuum theories	"
	37. Some aspects of ring theory	"
	38. Mathematical optimization in economics	"
1966 -	39. Calculus of variations	Ed. Cremonese, Firenze
	40. Economia matematica	"
	41. Classi caratteristiche e questioni connesse	"
	42. Some aspects of diffusion theory	"
1967 -	43. Modern questions of celestial mechanics	"
	44. Numerical analysis of partial differential equations	"
	45. Geometry of homogeneous bounded domains	"
1968 -	46. Controllability and observability	"
	47. Pseudo-differential operators	"
	48. Aspects of mathematical logic	"
1969 -	49. Potential theory	"
	50. Non-linear continuum theories in mechanics and physics and their applications	"
	51. Questions of algebraic varieties	"
1970 -	52. Relativistic fluid dynamics	"
	53. Theory of group representations and Fourier analysis	"
	54. Functional equations and inequalities	"
	55. Problems in non-linear analysis	"
1971 -	56. Stereodynamics	"
	57. Constructive aspects of functional analysis (2 vol.)	"
	58. Categories and commutative algebra	"
1972 -	59. Non-linear mechanics	"
	60. Finite geometric structures and their applications	"
	61. Geometric measure theory and minimal surfaces	"
1973 -	62. Complex analysis	"
	63. New variational techniques in mathematical physics	"
	64. Spectral analysis	"

1974 -	65. Stability problems	Ed. Cremonese, Firenze
	66. Singularities of analytic spaces	"
	67. Eigenvalues of non linear problems	"
1975 -	68. Theoretical computer sciences	"
	69. Model theory and applications	"
	70. Differential operators and manifolds	"
1976 -	71. Statistical Mechanics	Ed. Liguori, Napoli
	72. Hyperbolicity	"
	73. Differential topology	"
1977 -	74. Materials with memory	"
	75. Pseudodifferential operators with applications	"
	76. Algebraic surfaces	"
1978 -	77. Stochastic differential equations	"
	78. Dynamical systems	Ed. Liguori, Napoli and Birkhäuser Verlag
1979 -	79. Recursion theory and computational complexity	Ed. Liguori, Napoli
	80. Mathematics of biology	"
1980 -	81. Wave propagation	"
	82. Harmonic analysis and group representations	"
	83. Matroid theory and its applications	"
1981 -	84. Kinetic Theories and the Boltzmann Equation	(LNM 1048) Springer-Verlag
	85. Algebraic Threefolds	(LNM 947) "
	86. Nonlinear Filtering and Stochastic Control	(LNM 972) "
1982 -	87. Invariant Theory	(LNM 996) "
	88. Thermodynamics and Constitutive Equations	(LN Physics 228) "
	89. Fluid Dynamics	(LNM 1047) "
1983 -	90. Complete Intersections	(LNM 1092) "
	91. Bifurcation Theory and Applications	(LNM 1057) "
	92. Numerical Methods in Fluid Dynamics	(LNM 1127) "
1984	93. Harmonic Mappings and Minimal Immersions	(LNM 1161) "
	94. Schrödinger Operators	(LNM 1159) "
	95. Buildings and the Geometry of Diagrams	(LNM 1181) "
1985 -	96. Probability and Analysis	(LNM 1206) "
	97. Some Problems in Nonlinear Diffusion	(LNM 1224) "
	98. Theory of Moduli	to appear "

Note: Volumes 1 to 38 are out of print. A few copies of volumes 23,28,31,32,33,34,36,38 are available on request from C.I.M.E.

1986 -	99. Inverse Problems	(LNM 1225) Springer-Verlag
	100. Mathematical Economics	to appear "
	101. Combinatorial Optimization	to appear "

Vol. 1117: D.J. Aldous, J.A. Ibragimov, J. Jacod, Ecole d'Été de Probabilités de Saint-Flour XIII – 1983. Édité par P.L. Hennequin. IX, 409 pages. 1985.

Vol. 1118: Grossissements de filtrations: exemples et applications. Seminaire, 1982/83. Edité par Th. Jeulin et M. Yor. V, 315 pages. 1985.

Vol. 1119: Recent Mathematical Methods in Dynamic Programming. Proceedings, 1984. Edited by I. Capuzzo Dolcetta, W.H. Fleming and T. Zolezzi. VI, 202 pages. 1985.

Vol. 1120: K. Jarosz, Perturbations of Banach Algebras. V, 118 pages. 1985.

Vol. 1121: Singularities and Constructive Methods for Their Treatment. Proceedings, 1983. Edited by P. Grisvard, W. Wendland and J.R. Whiteman. IX, 346 pages. 1985.

Vol. 1122: Number Theory. Proceedings, 1984. Edited by K. Alladi. VII, 217 pages. 1985.

Vol. 1123: Séminaire de Probabilités XIX 1983/84. Proceedings. Edité par J. Azéma et M. Yor. IV, 504 pages. 1985.

Vol. 1124: Algebraic Geometry, Sitges (Barcelona) 1983. Proceedings. Edited by E. Casas-Alvero, G.E. Welters and S. Xambó-Descamps. XI, 416 pages. 1985.

Vol. 1125: Dynamical Systems and Bifurcations. Proceedings, 1984. Edited by B.L.J. Braaksma, H.W. Broer and F. Takens. V, 129 pages. 1985.

Vol. 1126: Algebraic and Geometric Topology. Proceedings, 1983. Edited by A. Ranicki, N. Levitt and F. Quinn. V, 423 pages. 1985.

Vol. 1127: Numerical Methods in Fluid Dynamics. Seminar. Edited by F. Brezzi, VII, 333 pages. 1985.

Vol. 1128: J. Elschner, Singular Ordinary Differential Operators and Pseudodifferential Equations. 200 pages. 1985.

Vol. 1129: Numerical Analysis, Lancaster 1984. Proceedings. Edited by P.R. Turner. XIV, 179 pages. 1985.

Vol. 1130: Methods in Mathematical Logic. Proceedings, 1983. Edited by C.A. Di Prisco. VII, 407 pages. 1985.

Vol. 1131: K. Sundaresan, S. Swaminathan, Geometry and Nonlinear Analysis in Banach Spaces. III, 116 pages. 1985.

Vol. 1132: Operator Algebras and their Connections with Topology and Ergodic Theory. Proceedings, 1983. Edited by H. Araki, C.C. Moore, Ş. Strătilă and C. Voiculescu. VI, 594 pages. 1985.

Vol. 1133: K. C. Kiwiel, Methods of Descent for Nondifferentiable Optimization. VI, 362 pages. 1985.

Vol. 1134: G.P. Galdi, S. Rionero, Weighted Energy Methods in Fluid Dynamics and Elasticity. VII, 126 pages. 1985.

Vol. 1135: Number Theory, New York 1983–84. Seminar. Edited by D.V. Chudnovsky, G.V. Chudnovsky, H. Cohn and M.B. Nathanson. V, 283 pages. 1985.

Vol. 1136: Quantum Probability and Applications II. Proceedings, 1984. Edited by L. Accardi and W. von Waldenfels. VI, 534 pages. 1985.

Vol. 1137: Xiao G., Surfaces fibrées en courbes de genre deux. IX, 103 pages. 1985.

Vol. 1138: A. Ocneanu, Actions of Discrete Amenable Groups on von Neumann Algebras. V, 115 pages. 1985.

Vol. 1139: Differential Geometric Methods in Mathematical Physics. Proceedings, 1983. Edited by H. D. Doebner and J. D. Hennig. VI, 337 pages. 1985.

Vol. 1140: S. Donkin, Rational Representations of Algebraic Groups. VII, 254 pages. 1985.

Vol. 1141: Recursion Theory Week. Proceedings, 1984. Edited by H.-D. Ebbinghaus, G.H. Müller and G.E. Sacks. IX, 418 pages. 1985.

Vol. 1142: Orders and their Applications. Proceedings, 1984. Edited by I. Reiner and K. W. Roggenkamp. X, 306 pages. 1985.

Vol. 1143: A. Krieg, Modular Forms on Half-Spaces of Quaternions. XIII, 203 pages. 1985.

Vol. 1144: Knot Theory and Manifolds. Proceedings, 1983. Edited by D. Rolfsen. V, 163 pages. 1985.

Vol. 1145: G. Winkler, Choquet Order and Simplices. VI, 143 pages. 1985.

Vol. 1146: Séminaire d'Algèbre Paul Dubreil et Marie-Paule Malliavin. Proceedings, 1983–1984. Edité par M.-P. Malliavin. IV, 420 pages. 1985.

Vol. 1147: M. Wschebor, Surfaces Aléatoires. VII, 111 pages. 1985.

Vol. 1148: Mark A. Kon, Probability Distributions in Quantum Statistical Mechanics. V, 121 pages. 1985.

Vol. 1149: Universal Algebra and Lattice Theory. Proceedings, 1984. Edited by S. D. Comer. VI, 282 pages. 1985.

Vol. 1150: B. Kawohl, Rearrangements and Convexity of Level Sets in PDE. V, 136 pages. 1985.

Vol 1151: Ordinary and Partial Differential Equations. Proceedings, 1984. Edited by B.D. Sleeman and R.J. Jarvis. XIV, 357 pages. 1985.

Vol. 1152: H. Widom, Asymptotic Expansions for Pseudodifferential Operators on Bounded Domains. V, 150 pages. 1985.

Vol. 1153: Probability in Banach Spaces V. Proceedings, 1984. Edited by A. Beck, R. Dudley, M. Hahn, J. Kuelbs and M. Marcus. VI, 457 pages. 1985.

Vol. 1154: D.S. Naidu, A.K. Rao, Singular Pertubation Analysis of Discrete Control Systems. IX, 195 pages. 1985.

Vol. 1155: Stability Problems for Stochastic Models. Proceedings, 1984. Edited by V.V. Kalashnikov and V.M. Zolotarev. VI, 447 pages. 1985.

Vol. 1156: Global Differential Geometry and Global Analysis 1984. Proceedings, 1984. Edited by D. Ferus, R.B. Gardner, S. Helgason and U. Simon. V, 339 pages. 1985.

Vol. 1157: H. Levine, Classifying Immersions into $\mathbb{R}^4$ over Stable Maps of 3-Manifolds into $\mathbb{R}^2$. V, 163 pages. 1985.

Vol. 1158: Stochastic Processes – Mathematics and Physics. Proceedings, 1984. Edited by S. Albeverio, Ph. Blanchard and L. Streit. VI, 230 pages. 1986.

Vol. 1159: Schrödinger Operators, Como 1984. Seminar. Edited by S. Graffi. VIII, 272 pages. 1986.

Vol. 1160: J.-C. van der Meer, The Hamiltonian Hopf Bifurcation. VI, 115 pages. 1985.

Vol. 1161: Harmonic Mappings and Minimal Immersions, Montecatini 1984. Seminar. Edited by E. Giusti. VII, 285 pages. 1985.

Vol. 1162: S.J.L. van Eijndhoven, J. de Graaf, Trajectory Spaces, Generalized Functions and Unbounded Operators. IV, 272 pages. 1985.

Vol. 1163: Iteration Theory and its Functional Equations. Proceedings, 1984. Edited by R. Liedl, L. Reich and Gy. Targonski. VIII, 231 pages. 1985.

Vol. 1164: M. Meschiari, J.H. Rawnsley, S. Salamon, Geometry Seminar "Luigi Bianchi" II – 1984. Edited by E. Vesentini. VI, 224 pages. 1985.

Vol. 1165: Seminar on Deformations. Proceedings, 1982/84. Edited by J. Ławrynowicz. IX, 331 pages. 1985.

Vol. 1166: Banach Spaces. Proceedings, 1984. Edited by N. Kalton and E. Saab. VI, 199 pages. 1985.

Vol. 1167: Geometry and Topology. Proceedings, 1983–84. Edited by J. Alexander and J. Harer. VI, 292 pages. 1985.

Vol. 1168: S.S. Agaian, Hadamard Matrices and their Applications. III, 227 pages. 1985.

Vol. 1169: W.A. Light, E.W. Cheney, Approximation Theory in Tensor Product Spaces. VII, 157 pages. 1985.

Vol. 1170: B.S. Thomson, Real Functions. VII, 229 pages. 1985.

Vol. 1171: Polynômes Orthogonaux et Applications. Proceedings 1984. Edité par C. Brezinski, A. Draux, A.P. Magnus, P. Maroni et A Ronveaux. XXXVII, 584 pages. 1985.

Vol. 1172: Algebraic Topology, Göttingen 1984. Proceedings. Edited by L. Smith. VI, 209 pages. 1985.

Vol. 1173: H. Delfs, M. Knebusch, Locally Semialgebraic Spaces. XVI, 329 pages. 1985.

Vol. 1174: Categories in Continuum Physics, Buffalo 1982. Seminar. Edited by F.W. Lawvere and S.H. Schanuel. V, 126 pages. 1986.

Vol. 1175: K. Mathiak, Valuations of Skew Fields and Projective Hjelmslev Spaces. VII, 116pages. 1986.

Vol. 1176: R.R. Bruner, J.P. May, J.E. McClure, M. Steinberger, H_∞ Ring Spectra and their Applications. VII, 388 pages. 1986.

Vol. 1177: Representation Theory I. Finite Dimensional Algebras. Proceedings, 1984. Edited by V. Dlab, P. Gabriel and G. Michler. XV, 340 pages. 1986.

Vol. 1178: Representation Theory II. Groups and Orders. Proceedings, 1984. Edited by V. Dlab, P. Gabriel and G. Michler. XV, 370 pages. 1986.

Vol. 1179: Shi J.-Y. The Kazhdan-Lusztig Cells in Certain Affine Weyl Groups. X, 307 pages. 1986.

Vol. 1180: R. Carmona, H. Kesten, J.B. Walsh, École d'Été de Probabilités de Saint-Flour XIV – 1984. Édité par P.L. Hennequin. X, 438 pages. 1986.

Vol. 1181: Buildings and the Geometry of Diagrams, Como 1984. Seminar. Edited by L. Rosati. VII, 277 pages. 1986.

Vol. 1182: S. Shelah, Around Classification Theory of Models. VII, 279 pages. 1986.

Vol. 1183: Algebra, Algebraic Topology and their Interactions. Proceedings, 1983. Edited by J.-E. Roos. XI, 396 pages. 1986.

Vol. 1184: W. Arendt, A. Grabosch, G. Greiner, U. Groh, H.P. Lotz, U. Moustakas, R. Nagel, F. Neubrander, U. Schlotterbeck, One-parameter Semigroups of Positive Operators. Edited by R. Nagel. X, 460 pages. 1986.

Vol. 1185: Group Theory, Beijing 1984. Proceedings. Edited by Tuan H.F. V, 403 pages. 1986.

Vol. 1186: Lyapunov Exponents. Proceedings, 1984. Edited by L. Arnold and V. Wihstutz. VI, 374 pages. 1986.

Vol. 1187: Y. Diers, Categories of Boolean Sheaves of Simple Algebras. VI, 168 pages. 1986.

Vol. 1188: Fonctions de Plusieurs Variables Complexes V. Séminaire, 1979–85. Edité par François Norguet. VI, 306 pages. 1986.

Vol. 1189: J. Lukeš, J. Malý, L. Zajíček, Fine Topology Methods in Real Analysis and Potential Theory. X, 472 pages. 1986.

Vol. 1190: Optimization and Related Fields. Proceedings, 1984. Edited by R. Conti, E. De Giorgi and F. Giannessi. VIII, 419 pages. 1986.

Vol. 1191: A.R. Its, V.Yu. Novokshenov, The Isomonodromic Deformation Method in the Theory of Painlevé Equations. IV, 313 pages. 1986.

Vol. 1192: Equadiff 6. Proceedings, 1985. Edited by J. Vosmansky and M. Zlámal. XXIII, 404 pages. 1986.

Vol. 1193: Geometrical and Statistical Aspects of Probability in Banach Spaces. Proceedings, 1985. Edited by X. Fernique, B. Heinkel, M.B. Marcus and P.A. Meyer. IV, 128 pages. 1986.

Vol. 1194: Complex Analysis and Algebraic Geometry. Proceedings, 1985. Edited by H. Grauert. VI, 235 pages. 1986.

Vol. 1195: J.M. Barbosa, A.G. Colares, Minimal Surfaces in $\mathbb{R}^3$. X, 124 pages. 1986.

Vol. 1196: E. Casas-Alvero, S. Xambó-Descamps, The Enumerative Theory of Conics after Halphen. IX, 130 pages. 1986.

Vol. 1197: Ring Theory. Proceedings, 1985. Edited by F.M.J. van Oystaeyen. V, 231 pages. 1986.

Vol. 1198: Séminaire d'Analyse, P. Lelong – P. Dolbeault – H. Skoda. Seminar 1983/84. X, 260 pages. 1986.

Vol. 1199: Analytic Theory of Continued Fractions II. Proceedings, 1985. Edited by W.J. Thron. VI, 299 pages. 1986.

Vol. 1200: V.D. Milman, G. Schechtman, Asymptotic Theory of Finite Dimensional Normed Spaces. With an Appendix by M. Gromov. VIII, 156 pages. 1986.

Vol. 1201: Curvature and Topology of Riemannian Manifolds. Proceedings, 1985. Edited by K. Shiohama, T. Sakai and T. Sunada. VII, 336 pages. 1986.

Vol. 1202: A. Dür, Möbius Functions, Incidence Algebras and Power Series Representations. XI, 134 pages. 1986.

Vol. 1203: Stochastic Processes and Their Applications. Proceedings, 1985. Edited by K. Itô and T. Hida. VI, 222 pages. 1986.

Vol. 1204: Séminaire de Probabilités XX, 1984/85. Proceedings. Edité par J. Azéma et M. Yor. V, 639 pages. 1986.

Vol. 1205: B.Z. Moroz, Analytic Arithmetic in Algebraic Number Fields. VII, 177 pages. 1986.

Vol. 1206: Probability and Analysis, Varenna (Como) 1985. Seminar. Edited by G. Letta and M. Pratelli. VIII, 280 pages. 1986.

Vol. 1207: P.H. Bérard, Spectral Geometry: Direct and Inverse Problems. With an Appendix by G. Besson. XIII, 272 pages. 1986.

Vol. 1208: S. Kaijser, J.W. Pelletier, Interpolation Functors and Duality. IV, 167 pages. 1986.

Vol. 1209: Differential Geometry, Peñíscola 1985. Proceedings. Edited by A.M. Naveira, A. Ferrández and F. Mascaró. VIII, 306 pages. 1986.

Vol. 1210: Probability Measures on Groups VIII. Proceedings, 1985. Edited by H. Heyer. X, 386 pages. 1986.

Vol. 1211: M.B. Sevryuk, Reversible Systems. V, 319 pages. 1986.

Vol. 1212: Stochastic Spatial Processes. Proceedings, 1984. Edited by P. Tautu. VIII, 311 pages. 1986.

Vol. 1213: L.G. Lewis, Jr., J.P. May, M. Steinberger, Equivariant Stable Homotopy Theory. IX, 538 pages. 1986.

Vol. 1214: Global Analysis – Studies and Applications II. Edited by Yu.G. Borisovich and Yu.E. Gliklikh. V, 275 pages. 1986.

Vol. 1215: Lectures in Probability and Statistics. Edited by G. del Pino and R. Rebolledo. V, 491 pages. 1986.

Vol. 1216: J. Kogan, Bifurcation of Extremals in Optimal Control. VIII, 106 pages. 1986.

Vol. 1217: Transformation Groups. Proceedings, 1985. Edited by S. Jackowski and K. Pawalowski. X, 396 pages. 1986.

Vol. 1218: Schrödinger Operators, Aarhus 1985. Seminar. Edited by E. Balslev. V, 222 pages. 1986.

Vol. 1219: R. Weissauer, Stabile Modulformen und Eisensteinreihen. III, 147 Seiten. 1986.

Vol. 1220: Séminaire d'Algèbre Paul Dubreil et Marie-Paule Malliavin. Proceedings, 1985. Edité par M.-P. Malliavin. IV, 200 pages. 1986.

Vol. 1221: Probability and Banach Spaces. Proceedings, 1985. Edited by J. Bastero and M. San Miguel. XI, 222 pages. 1986.

Vol. 1222: A. Katok, J.-M. Strelcyn, with the collaboration of F. Ledrappier and F. Przytycki, Invariant Manifolds, Entropy and Billiards; Smooth Maps with Singularities. VIII, 283 pages. 1986.

Vol. 1223: Differential Equations in Banach Spaces. Proceedings, 1985. Edited by A. Favini and E. Obrecht. VIII, 299 pages. 1986.

Vol. 1224: Nonlinear Diffusion Problems, Montecatini Terme 1985. Seminar. Edited by A. Fasano and M. Primicerio. VIII, 188 pages. 1986.

Vol. 1225: Inverse Problems, Montecatini Terme 1986. Seminar. Edited by G. Talenti. VIII, 204 pages. 1986.